UNFOLDMENT AND MANIFESTATION

UNFOLDMENT AND MANIFESTATION

Seven essays on evolution and classification

L. van der Hammen

SPB Academic Publishing 1988

The cover design symbolizes the deep structure underlying evolution; it was inspired by one of the medieval French rose windows which, as mentioned in essay I of this book, can be regarded as models of the unfolding universe.

© 1988 SPB Academic Publishing, P.O. Box 97747
2509 GC The Hague, The Netherlands

ISBN 90-5103-008-8

CIP-DATA KONINKLIJKE BIBLIOTHEEK, DEN HAAG

Hammen, L. van der

Unfoldment and manifestation : seven essays on evolution
and classification / L. van der Hammen. – Den Haag :
SPB Academic Publishing. – Ill.
With bibliogr.
ISBN 90-5103-008-8
SISO 574 UDC 575.8 NUGI 821
Subject heading: evolution : essays.

TABLE OF CONTENTS

INTRODUCTION

The essays collected in the present book contain elements of a new theory of evolution and classification. They were written in a period of about nine years, and proceed from theoretical problems which arose during my studies of the comparative morphology, the classification and the evolution of Chelicerata. The theoretical considerations finally led to a structuralist approach in the study of these subjects.

It may be remarked here that the results of my studies in chelicerate evolution had, from the outset, been at variance with the current neo-Darwinian views of evolution (according to which mutations and recombinations in populations constitute the sole material for natural selection). Besides that, current taxonomic methods appeared to be insufficient for the solution of problems with reference to the higher, i.e. supraordinal, classification of Chelicerata (the current methods tend indeed to develop atomistic views in which the organism as a functional whole has virtually no place, whilst the hierarchic value of characters is, generally, insufficiently analysed).

As mentioned in essay VI, the choice of Chelicerata as the subject of my studies appeared to have several interesting aspects for theoretical considerations. Chelicerata are segmented animals, of which the postembryonic development is characterized by moulting and the occurrence of a series of separate instars; besides that, parthenogenesis is found in several groups of mites. The origin of segmentation (a key event in the evolution of many phyla) must be partly connected with a repeated manifestation of the same genetic pattern; it is often followed, in the course of evolution, by the superimposed manifestation of divergency (probably as a result of specific genetic and developmental interactions). A postembryonic development which is characterized by moulting, has the advantage that a restricted (and often fixed) number of levels of transformation can be distinguished (the manifestation of changes is concentrated at distinctly separated levels, by which an analysis is facilitated). The occurrence of parthenogenesis, finally, has the advantage that the manifestation of certain variations can be analysed not only in bisexual populations, but also in clones; in the case of the present research, this analysis has led to the concept of evolutionary potentiality and the concept of probability of manifestation. For several years, the theoretical studies were carried out simultaneously with my studies in Chelicerata and, in writing the two series of studies, insights developed and influenced each other mutually. Although the theoretical studies were started as a contribution to the studies in Chelicerata, the last-mentioned studies were finally also adapted to contribute to the solution of theoretical problems.

My development as an arachnologist has been considerably influenced by my close collaboration, for a period of more than twenty years, with the French acarologist F. Grandjean. Many data on Grandjean's life and works, and on my indebtedness to him, can be found in the introductions to my edition of his complete acarological works in seven volumes (F. Grandjean, Œuvres acarologiques complètes, The Hague & Lochem, 1972–1976). In the course of my collaboration with him, I learned to understand the importance of careful and detailed observations, the importance of the comparative study of the postembryonic development for the study of evolution, the importance of a standardization of methods, and the importance of the development of a uniform terminology (which led to my editing of the Glossary of Acarological Terminology, The Hague, 1976, 1980).

Important influences on the development of the theoretical essays are also constituted by my interest in classical, mediaeval and Renaissance philosophy, and by my interest in the literature and natural philosophy of Goethe and his contemporaries. The final presentation of the essays in this book owes much to my participation in the International Workshop on Structuralism in Biology (Osaka, Japan, December 1986), and to its organizer Atuhiro Sibatani, who invited me. During this workshop, I had the opportunity to meet specialists in related disciplines, particularly cytogenetics and morphogenetics, which disciplines have important points of contact with my own fields of study.

My theoretical studies brought me indeed repeatedly into contact with subjects in the fields of other disciplines, with which I am not at home (this is, in fact, the only way to free an investigation from the limitations of a specialism). Along this line, I have tried to explore new approaches and to make the results generally accessible. In doing so, I have repeatedly been obliged to correct myself, when it appeared that I was on the wrong track. In the course of writing the essays, it has indeed become evident that my studies have points of contact with modern research in the fields of molecular cytogenetics (A. Lima-de-Faria) and morphogenetics (C.H. Waddington and his school, A. García-Bellido, R. Thom), and with modern abstract coding theory approaches (S.A. Kauffman, D. Elder); besides that, there are, for instance, also points of contact with earlier studies by D'Arcy W. Thompson, and with the natural philosophy of the theoretical physicist D. Bohm. It goes without saying that the following essays contain only the first elements of new approaches, and the first beginnings of new theories; there is still no question of a coherent whole. The essays are, in fact, more or less disconnected, and sometimes repeat the same things in a different context.

It has now become evident that the future of research on evolution lies with comparative ontogenetic studies, and that partial explanations are to be found in the field of changes in epigenetic coding, which changes can be co-ordinated with changes in the environment (the ultimate nature of the co-ordination is really a metaphysical problem). Further explanations will be in the fields of mathematics and natural philosophy. In my opinion, the future of systematics lies with a new kind of numerical taxonomy (structuralist taxonomy), that must be founded on a hierarchical model of evolution (ultimately perhaps a mathematical structure).

In the first essay, a study is made of the general principles and theories with reference to evolution, among which the definition, the evidences, the philosophical roots (theories on the origin of life, the scale of nature, and the morphogenetic potentialities), and the three models (the Lamarckian, the neo-Darwinian and a third model); in the final section of the essay the third model, in which evolution is attributed to a co-ordination of internal and external factors, is further developed (it deals with aspects of manifestation, gradual unfoldment and the stages of evolution, and systems of interactions, vital field, and the implicate order).

In the second essay, a survey is given of the evolutionary aspects of numerical changes in Actinotrichid mites. Numerical variations are studied in clones, in bisexual populations and in supraspecific taxa. Special attention is paid to the ontogenetic aspects of the manifestation of evolution, and to the relative strengths and weaknesses in series of homonomous elements. The numerical changes are associated here with changes in systems of genotypic interactions (evolutionary changes in epigenetic coding); in one case, there is evidence that these changes are co-ordinated with changes in the environment. It is pointed out that some of the properties of the evolutionary code underlying the epigenetic code (there is an evolutionary potentiality and a probability of manifestation) could perhaps give an indication of the nature of the code. A general survey is, moreover, given of the main numerical changes in the representatives of one superfamily; it is suggested that continued research in this field could lead to the preparation of a complete mathematical model of the evolution of chaetotaxy in this group.

Essay III, devoted to the chelicerate life-cycle, deals with forms and moults, phases of similar instars, regressive forms, and evolutionary phenomena. The successive forms of the life-cycle are regarded as developmental variations, the manifestation of unfoldment in the course of time, of a compound genetic pattern (itself a spatial series of variations of a single genetic pattern); the evolution of this life-cycle is regarded as changes in these developmental variations. A general model is prepared for the evolution of the life-cycle.

In essay IV, a study is made of some aspects of parallel evolution in Chelicerata. Definitions are given of parallel evolution, convergence, homology and analogy. It is pointed out that the concept of parallel evolution (parallelism) is initially formed in an empirical way, and that a subsequent judgment must be based on certain criteria. Particular attention is paid to the role of gene regulation in parallel evolution, to the special case of convergence as a result of heterologous regulatory mechanisms, to parallel evolution in homonomous structures (and the superposition of parallelisms and divergencies), and to parallelism in the evolution of characters used in higher classification.

In essay V, a study is made of the history of the type and related concepts, from Greek Antiquity up to the present. It is demonstrated that the type-concept of eighteenth-century biology was based on Leibniz's concept of substantial form, whilst it is now generally understood in the sense of model or norm. It is pointed out that one of the essential characters of the eighteenth- and nineteenth-century type-concept was constituted by its being composed of homonomous parts (homotypes); according to recent discoveries in morphogenetics and molecular cytogenetics, these must now be regarded as developmental variations of a single basic genetic pattern. In the essay, a type-concept is developed which can serve as a model of the evolutionary potentialities of a taxon, and as a standard of (particularly higher) classification; this model has a hierarchical structure, and includes ontogenetic and phylogenetic time, and various evolutionary mechanisms.

In essay VI, a survey is given of structuralism as a method that can be applied in the study of evolution and classification. The results of a structuralist approach are illustrated by examples (based on essays II, III and IV) from the laws underlying numerical changes, from the laws underlying changes in the chelicerate life-cycle, and from the laws underlying the evolution of the chelicerate appendages; suggestions are given for the study of the laws underlying transformation of form. Some concluding remarks are made on deep structure and classification.

In the last essay, some introductory notes are made on taxonomic methodology. After an analysis of taxonomic practice, and a brief survey of kinds of attributes, the essay deals with observation, description, comparison, arrangement and classification, hypothesis construction, deduction, model, experiment, abstraction and synthesis. The methodological aspects of the species-concept and of biological classification are dealt with in two final sections.

A first application of the results of the essays can be found in my comparative studies in Chelicerata. These studies are now in course of being summarized in a book entitled *An Introduction to Comparative Arachnology*, publication of which can be expected before long.

The contents of the essays could be the starting-point for a detailed research program (execution of which could lead to very interesting new results). The philosophical part of the essays could be extended by a more detailed study of the influence of Leibniz on the French morphologists and their natural philosophy, particularly by an analysis of the contemporary French periodical publications in the field of the sciences, such as the Journal des Savants (other periodicals in this field are, e.g., mentioned by D. Mornet, Les Sciences de la Nature en France, au XVIIIe Siècle. Un chapitre de l'Histoire des Idées, Paris, 1911). A structuralist reinterpretation of Driesch's vitalism in the light of Thom's vital field (a multidimensional geometrical structure; see: R. Thom,

Stabilité structurelle et Morphogénèse, Reading, Mass., 1972), will also be interesting; this reinterpretation could perhaps include a critical discussion of the studies by Conrad-Martius on the entelechy and related concepts (Der Selbstaufbau der Natur: Entelechien und Energien, Hamburg, 1944; Die Geistseele des Menschen, München, 1960).

The data published in the second essay could be the starting-point for the preparation of a model of the hierarchical structure underlying the postembryonic development and the evolution of chaetotaxy in a group, e.g., in the Oribatid superfamily Nothroidea (a model in which epigenetic and evolutionary coding are clearly distinguished). It would also be interesting to extend the study of the claws in *Ameronothrus schneideri* (an example of an evolution based on the co-ordination of internal and external factors), by the analysis of material from additional localities, and by a detailed comparative study of the environment.

A careful analysis of evolution in a taxon of large content could lead to a better understanding of the hierarchic development of new evolutionary potentialities, and to a discrimination between that development and the manifestation of transmutation by changes in epigenetic coding. This study should include an analysis of the morphologies that do not occur in nature (the 'hiatotypes' of essay V). The important problem of transformation of form, in the course of evolution, could be studied by the detailed analysis of shape in a number of closely related species, by application of D'Arcy Thompson's method of co-ordinates (see: Thompson, On Growth and Form, Cambridge, 1917).

As mentioned above, the final presentation of the essays included in this book, owes much to the International Workshop on Structuralism in Biology, in Osaka, Japan, 1986. For this reason, I should like to thank its organizer Atuhiro Sibatani and all the participants for the many stimulating communications and discussions. I am also greatly indebted to David Elder for his interest and encouragement, and for reading and commenting on parts of the typescript.

Baronielaan 1
2242 RA Wassenaar

L. van der Hammen
August 1987

I. UNFOLDMENT AND MANIFESTATION: THE NATURAL PHILOSOPHY OF EVOLUTION

1. Introduction

Four years ago I published, in Acta Biotheoretica, a paper under the same title as the present essay (Van der Hammen, 1983); it constituted the philosophical introduction to a series of studies on evolution and classification written not long before (Van der Hammen, 1978, 1981, 1981a). It is now entirely revised, enlarged and partly rewritten to serve, as essay I, the same introductory purpose in the present book. The essay includes a definition of the concept of evolution, a general survey of the evidences, and a study of the philosophical roots of the idea of evolution (speculations with reference to the origin of life, the scale of nature, and the morphogenetic potentialities). Thereupon, three models of evolution are discussed: the Lamarckian model, the Darwinian model, and a model in which evolution is attributed to a co-ordination of internal and external factors. In the last section of the present essay, the third model is further developed.

2. The term 'evolution' and its definition

The English word 'evolution' is derived from the Latin *evolutio*, which pertains to unrolling (particularly of a scroll). In the nineteenth century the term had the general meaning of gradual change and, besides that, the biological meaning of embryonic development. Lyell, in the first (1832) edition of his *Principles of Geology*, used the term incidentally in his discussion of Lamarck's ideas on transmutation (cf. Lyell, 1872, vol. 2: 256). In *The Origin of Species*, Darwin used the term exceptionally, and particularly to indicate the slow and gradual character of transmutation (cf. Darwin, 1872: 201, 202, 424). Th.H. Huxley, in 1878, was the first to use the term in the modern sense of organic evolution. He extended the original biological meaning of evolution (as individual, i.e. embryonic and postembryonic development) to the development of the sum of all living beings (cf. Th.H. Huxley, 1893: 187–226).

There are several modern definitions of evolution, but the current one is neo-Darwinian and refers to the transformation of the genetic composition of populations, by interaction with the environment (cf. Dobzhansky, Ayala,

2

Stebbins & Valentine, 1977: 7–9). In a philosophical study of the phenomenon of evolution, which includes a discussion of the main models, such a narrow definition is unsuitable as a starting-point. For this reason, organic evolution is defined here as a process of transmutation in nature, including the origin of life, the genesis of the different types of organisms, and the change of these in the course of time.

3. Evidences of evolution

Direct, experimental proofs, i.e. conclusive evidences of evolution, are only present in the field of microevolution (at the infraspecific level). Indirect proofs, i.e. evidences which are not conclusive (and can be interpreted in different ways) are numerous and can be arranged under a number of headings, of which the following are mentioned here.

3.1. *Paleontological evidences*

The fossil record demonstrates that the simplest living beings occurred first, and that the more complicated arose afterwards: the age of the oldest known prokaryote fossils (bacteria) is estimated at 3.3 billion years, that of the oldest known eukaryote fossils at 1.3–1.7 billion years, that of the first Metazoa at 1.0 billion years (cf. Dobzhansky, Ayala, Stebbins & Valentine, 1977: 377, 397). Splendid examples of the succession of fossil forms have been studied by paleontologists. Many phylogenetic trees are based on a circumstantial fossil evidence. The detailed study of single lineages demonstrates, however, that there are always gaps; these must either be attributed to macroevolution (i.e. large-scale evolution owing to important changes in the genotype), or explained by missing links. Important paleontological evidences are constituted by extinct groups that have an intermediate position between extant groups. An interesting problem is constituted by the sudden appearance of the phyla, in the course of geological time, and the presence, from the beginning, of important discontinuities separating these. Another interesting problem is constituted by the fact that the simplest forms have continued to exist.

3.2. *Systematic evidences*

The gradation in living nature, extending from simplicity to very great complexity, is one of the foundations of biological systematics. Bacteria and blue-green algae are among the simplest living organisms, Angiospermae and Vertebrata among the most complicated. The natural system reflects, to some extent, the increasing amount of complexity, which is supposed to have arisen in the course of evolution. Among the representatives of natural groups various degrees of affinity are more or less evident; these can be interpreted in terms of common descent.

3.3. *Morphological evidences*

Important morphological evidence is constituted by the unity of plan of construction in a natural group, which points to a common origin. Interesting evidences result from the study of vestigial structures and atavisms, and from ecological morphology (e.g. the comparative study of related species with different habitat).

3.4. *Embryological evidences*

The existence of a certain analogy between embryological development and the gradation of living organisms has sometimes been regarded as evidence of evolution (this so-called biogenetic law is now more or less obsolete). Important evidence is constituted by the temporary occurrence, in the course of embryonic development, of structures that are no more necessary in the postembryonic stage (e.g. gill-slits in Mammals). It is often not taken into account that, in the case of evolution, the individual development (embryonic and postembryonic) must have evolved as well (evolutionary changes could have occurred at all levels of development).

3.5. *Zoogeographic evidences*

Facts pertaining to geographic distribution can be explained in an evolutionary sense, such as the occurrence of geographical races of one species, and the occurrence of related species on separate islands. Discontinuous distribution patterns of supraspecific taxa can also be interpreted in an evolutionary sense.

3.6. *Biochemical evidences*

Various data from biochemistry, such as the structure of proteins, the universality of cell structure and the basic structure of genetic material point to the fundamental unity of life, and to a common origin. Similarities and differences in amino acid sequences of proteins are nowadays regarded as a measure of relationship.

4. Philosophical roots of evolutionary thought

Apart from mythology, magic and the like, the idea of biological transformation has been developed, in particular, by Leibniz, the early French 'evolutionists' and Charles Darwin. Leibniz had a profound knowledge of contemporary science (including paleontology). This enabled him in 1691, in his *Protogaea*, to formulate ideas pertaining to the transformation of species (cf. Glass, Temkin & Straus, 1959: 38, 233n; Harris, 1981: 104). Leibniz's influence

4

on the French evolutionists was partly through his monadology and the recognition of an internal, self-organizing principle in matter and living organisms (cf. Leibniz/Loemker, 1969: 643–644). Ideas on transformation are still vague and contradictory in Buffon (cf. Buffon & Daubenton, 1753: 379–381), more pronounced (although with reservations) in Diderot (1754; cf. Harris, 1981: 115–116), and finally distinctly formulated by Lamarck (1809: first part, chapter 7; cf. Harris, 1981: 116–120). The philosophical roots of evolutionary thought, however, are much older than the idea of transformation, and reach back to Antiquity. These roots are discussed here under three headings: (1) the origin of life; (2) the scale of nature; and (3) the morphogenetic potentialities.

4.1. *The origin of life*

For a very long time, the origin of life was regarded either as a creative act of God, as a mythological subject, or as a theme for primitive philosophical speculation. At the same time, it was generally believed that certain simple organisms originated spontaneously from matter (cf. Oparin, 1953: 1–28). Aristotle distinguished between the reproduction of one's kind and spontaneous generation (Aristotle, *De Anima* II, 415a, 28). Even Lamarck, in his *Philosophie zoologique* (Lamarck, 1809: second part, chapter 6; third part, additions), supposed that the organisms lowest in the scale of nature had arisen by spontaneous generation (he distinguished at least two branches of animals, of which the first representatives arose in this way).

The impossibility of the current views of spontaneous generation was at last definitely demonstrated by the investigations of Pasteur (1861). It is interesting to note that theories on the origin of life at some distant period of the earth's existence (cf. Oparin, 1953: 45–63) arose at about the same time as Pasteur's refutation of spontaneous generation.

It is now recognized that primordial life must have had two properties: on the one hand the capacity of reproduction and growth (the appearance of complex molecules capable of continuous self-perpetuation without diminution), on the other hand the capacity of harnessing energy (the appearance of complex molecules functioning as catalysts). Evidently, these completely different properties cannot have been simultaneously present in a single homogeneous substance. The theory of the origin of life, consequently, presupposes an association or the associated development of the two capacities (cf. Good, 1981: 115–158).

4.2. *The scale of nature*

The idea of a scale of nature, the recognition of a graduated, uninterrupted series of forms in organic life, extending from the simplest to the most perfect, appears first in Plato. Through the Middle-Ages, and down to the beginning of the nineteenth century, this conception of the universe as a 'Great Chain of

Being' (cf. Lovejoy, 1936) was accepted by most educated men.

In the *Timaeus*, the dialogue in which the current cosmology of Antiquity is summarized, Plato wrote that the Father of the Universe created the world-soul as something standing between Being and Becoming. It was God's purpose that the world should consist of the completeness of parts, and that nothing should remain from which another, similar being could arise. He created the gods (the children of the Father) and the souls, but, because the other living beings should be mortal, He created their souls only, whilst the gods (the children of the Father) wove the mortal beings with the immortal souls. Together with man, the gods created the plants, in aid of mankind. The animals, however, arose by degeneration of man; he, who had not made the best of his span of life, was changed, at the next birth, into a lower being. The whole perceptible world arose, in likeness to the Realm of Being, in something different (a formless, invisible matter, a Receptacle comparable with a mother, in which becoming is enacted) (cf. *Timaeus* 34c−35a, 32d, 33a, 41a−d, 77a, 91d−92c, 48e−51a).

In these passages, several ideas are expressed which occupied science for a long time. Plato first mentioned the fullness of creation: all possible living beings had been realized. In the second place he mentioned that animals arose by degeneration of man. In the initial stage of zoology, this was a fruitful hypothesis: the functional meaning of animal organs could, in the first instance, only be understood in analogy to man.

Aristotle (*Historia animalium* VIII, 1, 588b; *De Partibus Animalium* IV, 5, 681a) developed Plato's principle of the completeness of Nature, and pointed out the phenomenon of continuity. Although classes can be distinguished, there are always transitional forms between the classes: transitional forms between inanimate nature and plants, transitional forms between plants and animals. The zoophytes (among which sea-anemones) constituted a favourite example of transitional forms between plants and animals. Bats took an intermediate position between animals living on the ground and flying animals; seals took an intermediate position between terrestrial and aquatic animals.

Neoplatonic and mediaeval philosophy added little to the views of Plato and Aristotle, as far as the biological aspect of the scale of nature is concerned. In the course of the seventeenth century, with the work of Leibniz, an important development sets in. As his predecessors, Leibniz regarded fullness or plenitude, continuity and linear gradation as essential characteristics of the universe. According to him, the scale of nature consists of the totality of monads (cf. subsection 4.3, below), ranging from God to the lowest grade of organic life, each monad possessing an internal self-organizing power (cf. Lovejoy, 1936: 144−145; Leibniz/Loemker, 1969: 643−644).

Buffon still regarded the scale of nature as a single descending series. In the opening discourse of the first volume of his *Histoire naturelle* (cf. Buffon, 1749: 12, 13, 20, 38), he wrote that the course of nature is by gradations, and that it is always possible to descend by almost insensible degrees from the most perfect creature to formless matter.

6

A reversal of the scale of nature, from a descending to an ascending series, is found in the work of Robinet. He wrote several books on the Chain of Being (Robinet, 1761–1766, 1768), in which philosophy and natural history are mixed with absurdities (he believed in the existence of mermaids). He regarded the Chain of Being as a realization of a single prototype (cf. subsection 4.3). This realization took place in an ascending series, extending from matter to man; man was considered the most perfect realization of this prototype. The German philosopher Herder (1784) also recognized a single prototype manifesting itself with increasing perfection in an ascending series of forms.

Lamarck (1809), in his *Philosophie zoologique*, wrote that the animal chain extended from the most perfect to the most imperfect animals, but that one could pass through the animal scale in the opposite direction, which is according to the order of nature. In that direction, one finds an increasing complexity in the animal organization (Lamarck, 1809: first part, chapter 6). In the additions to the third part of the book, and pertaining to chapters 7 and 8 of the first part, he dealt also with the origin of living animals; he arrived at the conclusion that the animal scale commences with at least two branches (Vermes and Infusoria). In Lamarck's work, the single animal chain has given way to two or more separate chains.

4.3. *The morphogenetic potentialities*

The concept of form, and the theory of form and matter, as developed in Antiquity, are closely connected with problems (at that time not yet understood) pertaining to food-intake, growth, metamorphosis and reproduction, i.e. to the repeated return of the same shapes. According to Plato (as mentioned already above, in subsection 4.2.), the whole perceptible world arose, in likeness to the Realm of Being, in something different. He described this as a formless, invisible matter, a Receptacle comparable with a mother, in which becoming is enacted. Form is, in Plato, apparently identical with soul.

Aristotle developed the natural philosophy of his teacher (*Physica* II, 1, 193b, 4–8, 19–20; *Metaphysica* VII, 2, 1028b, 3–5). Claghorn (1954: 134–135) has demonstrated that Aristotle's agreement with Plato is far greater than his divergence, and he quoted Jaeger who characterized Aristotle's criticisms of Plato's natural philosophy as arising out of the greater nearness to Plato, and as being essentially discussions within the Academy. Plato's Receptacle appears to correspond with Aristotle's prime matter (*hyle*) which is also formless and non-being. Plato's 'soul' has, in Aristotle's natural philosophy, become an immanent form (*eidos, morphe*). Every body (inorganic as well as organic) requires a form, but every living body possesses a soul which, in this case, is the particular name of its form. Aristotle's *eidos* is at the same time the lowest taxonomic category in the hierarchic scale of classification: different individuals of the same species have the same *eidos*.

Aristotle's term entelechy (*entelecheia*), which was probably coined by him, is connected with his theory of *energeia* (actualization of realization; *actus* in Scholastic philosophy) and *dynamis* (potentiality; *potentia* in Scholastic philosophy); it pertains to the actualization of a potentiality (the condition of completion, which includes a new potentiality). The soul is the first entelechy (first actuality) of a natural body (Aristotle refers here to the possession of a capacity, as opposed to its exercise).

Mediaeval Scholasticism adopted the concepts of form and matter, *energeia (actus)* and *dynamis (potentia)* from the Aristotelian philosophy. Thomas Aquinas distinguished, moreover, between substantial form (pertaining to Being) and accidental form (in which accidents or non-essentials are predicated of it) (cf. De Vries, 1980).

The mediaeval concept of the seminal reasons or causes (*rationes seminales*) is closely connected with the theory of form and matter. The concept was first introduced (as *logoi spermatikoi*) in the Stoic philosophy, and subsequently mentioned by Philo Alexandrinus and Plotinus. Augustine adapted it to his Christian philosophy (Creation was accomplished all at once, and all together, and included germinal forces of all that subsequently developed and will be developed). The concept of the seminal reasons or causes constitutes an important part of the neoplatonic philosophy of the cathedral school of Chartres (the rose windows of its cathedrals can be regarded as geometrical models of the unfolding universe). According to this philosophy, God operated in four ways: (1) in the Word, by conceiving the form of all things; (2) in matter; (3) in the seminal causes (implanted in matter); and (4) in the succession of time (cf. Häring, 1956; Van der Meulen, 1966).

Partly as a result of his correspondence with Arnauld, Leibniz published in 1695, in the '*Journal des Savants*' (a well-established Paris journal), his new system of the nature of substances (Leibniz/Loemker, 1969: 453–459). In this paper he wrote that, in turning back to Aristotle (Leibniz had also a profound knowledge of Scholasticism), he perceived that it is impossible to find the principles of a true unity in matter alone. In this way he discovered (i.e. postulated) genuine unities, helped by the old notion of substantial forms; he conceived these as centres of force or active power. Substantial forms resemble souls, or what Aristotle calls 'first entelechies' (a substantial form is to any organism as the soul is to an organism endowed with awareness). In 1696 Leibniz adopted the term 'monad' in place of 'substantial form' (the last-mentioned term reappeared in Leibniz's *New Essay on Human Understanding*, completed in 1704, but first published in 1765). Monads confer a kind of individuality on organisms. Every body is a colony of monads, but the deeper unity which characterizes organisms consists in their possessing a dominant monad. In the *Monadology* (1714) Leibniz restricted the term 'soul' to entelechies possessing consciousness. In his later writings Leibniz uses 'entelechy' interchangeably with 'monad', although, according to Broad (1979: 88–89), Leibniz distinguished 'entelechy' from 'monad' when he was careful (in these cases he

8

regarded an entelechy as an inseparable factor in a monad). As mentioned above, Leibniz attributed to each monad a self-organizing internal power (cf. Broad, 1979: 75–124; Loemker, 1969: 35–37; Remnant & Bennett, 1981: xxxix-xl, lvii, lxxiv).

The development of the type-concept from the concept of substantial form is dealt with in my paper on *Type-concept, higher Classification and Evolution* (Van der Hammen, 1981; cf. essay V) in which I pointed out the key-position of Robinet (1761–1766, 1768). Robinet's fundamental element in nature (something in the shape of a hollow cylinder) reminds one of Leibniz's monad. Robinet's prototype reminds one of Leibniz's dominant monad. The great difference is constituted by the fact that this prototype is not the unitary substance of a single individual, but of the whole scale of nature.

Before Robinet, Buffon had regarded the prototype as the unitary substance of a species, whilst his 'dessein primitif et général' pertains to greater systematic units (cf. Buffon & Daubenton, 1753: 215–216, 379–381). His 'moules intérieurs' correspond more or less with the genotype-concept (cf Buffon, 1749a: 41–46).

It may be added here that the concept of entelechy reappears in the philosophy of Driesch (cf. Driesch, 1928: 285), where it is defined as the causal base of form and action. Conrad-Martius (1944, 1960), a disciple of Husserl, re-defined the entelechy-concept and differentiated between several related concepts by applying the phenomenological method.

4.4. *Summary*

It is demonstrated in the preceding subsections that the most important landmarks in the development of early evolutionary thought are constituted by: (1) the refutation of current views of spontaneous generation by Pasteur, and the introduction of a new theory on the origin of life at some distant period of the earth's existence; (2) the reversal of the scale of nature from a descending into an ascending scale (by Robinet), and the division of the scale into several chains (by Lamarck); and (3) the development of the concept of substantial form, applied to species, into a type-concept, applied to a group (by Robinet), and the recognition of an internal self-organizing principle in monads (by Leibniz).

5. Models of evolution

There are two current models of evolution: the Lamarckian and the Darwinian, of which the latter has been the most successful in the world of science. To these two models, a third can now be added, to which several authors have recently contributed, and which is further developed in the present essay; as mentioned below, this model was, in reality, already introduced by the philosopher Teichmüller (1877), and resulted from his discussions with the embryologist K.E.

von Baer. The three models are briefly discussed in the following subsections.

5.1. *The Lamarckian model*

Lamarck developed his transformation model in his *Philosophie zoologique* (Lamarck, 1809: first part, chapter 7). He pointed out the fact that the scale of nature is not gradual, but presents gaps. According to him, nature tends indeed to a regular gradation in the increasing complexity of its organization, but this regularity is disturbed by many influences. The influence of the environment on animal form and organization, however, is not direct but indirect. By changes in the environment, the necessities of life in animals, and, in connection with it, the actions and habits, are modified. Organs arise from new necessities of life and are gradually improved, or gradually disappear in the case of disuse. Characters, acquired in this way, are inherited by the descendants. In the Lamarckian model, modification in the organism as well as change in the environment have been taken into account, although the former is considered the most important. From the beginning of this century, several authors (among which Berg, 1926) have recognized again the existence of an internally directed evolution; different names (such as orthogenesis) have been applied to this inherent directionality. In his discussion of Croizat's 'orthogeny', Grehan (1984) has given a compendious survey of the relevant literature. Insights developed by these authors, can be included in the alternative model discussed in subsection 5.3.

5.2. *The Darwinian model*

In the original Darwinian model of 1859 the main elements are constituted by: hereditary variation, struggle for life, natural selection, and the preservation of favoured races. Evidently, Darwin emphasized an element which was of minor importance in the Lamarckian model. Whilst the latter could be characterized as the development of a tradition, the nature of Darwin's contribution to evolutionary biology was revolutionary. In subsequent editions of *The Origin of Species*, Darwin modified his theory, and particularly from the fifth and sixth edition the theory included also non-selective evolutionary forces (cf. Liepman, 1981, 1982).

The neo-Darwinian model (the synthetic theory of evolution) introduced by Fisher (1930) and further developed by J.S. Huxley (1942) and others, is based on population-environment interactions; random mutations and recombinations in populations constitute the material for natural selection; in the case of bisexual reproduction, reproductive isolating mechanisms are supposed to counteract the disintegration of a gene pool (the sum total of the genotypes of all individuals in a reproductive community). In the neo-Darwinian model environmental selection is again emphasized. Molecular cytogenetics (cf. Lima-de-Faria, 1983, 1986) has recently demonstrated that chromosomes evade randomness and selection (the new concept of the chromosome field implies order within the chromosome), by which the neo-Darwinian model has become of minor importance.

5.3. *The third model of evolution*

A third model of evolution was introduced, more than a century ago, by the philosopher Teichmüller (1877). It arose from his discussions with K.E. von Baer who, in the last years of his life, occupied himself with the study of Darwinism, and who had asked Teichmüller to make a philosophical study of the problem.

Von Baer believed in the transmutation of species. His main objection to Darwinism was that adaptation was supposed to have arisen by elimination of the unadapted, and that too much importance was attached to chance. Von Baer's own view of the development of nature was more or less teleological. He regarded evolution as something analogous to ontogeny: a rapid development (with important transmutations) at the beginning of evolution, a much slower and more gradual development later on. He regarded transmutation as a process manifesting itself in the course of ontogeny, and resulting in important modifications (cf. Von Baer, 1876: 49–105, 170–480; Stölzle, 1897: 195–289).

Teichmüller (1877) criticized Darwinism because it over-emphasized external causes. According to him, transformation can only take place by a co-ordination of internal and external forces, although the laws of transformation (and the evolutionary potentialities) are in the type (the normative form, i.e. the genotype). A change in the internal system influences all parts of the type. For this reason genuine transformation (which is different from a simple variation) is always saltatory. The origin of new types requires a co-ordinated change of the complete system; this change could be connected with the origin of new geological periods.

Recent developments in molecular biology, paleontology and cybernetics have contributed to the development of Teichmüller's model of evolution. Part of these developments were recently summarized by Van Waesberghe (1981, 1982; he was not acquainted with Teichmüller's model), and incorporated into his 'alternative evolution model'. He regarded evolution as the result of a co-ordination of external and internal factors (environment and genotype), and the course of evolution as slow or rapid, gradual or saltatory. He admitted micro- and macromutations, but regarded speciation as the result of macroevolution ('punctuational' change). Consequently, gaps in a complete fossil record are symptoms of macromutations. Saltatory evolution is provoked by a sudden collapse of evolutionary thresholds (external and internal). Van Waesberghe interpreted the genotype cybernetically as a program of action and reaction (not as a cluster of genes), a gene as a readily split genetic factor, and the DNA sequence as the material base of the gene.

6. Further development of the third model

As mentioned in the preceding section, evolution can be regarded as the result of a co-ordination of internal and external factors (genotype and environment), although the evolutionary potentialities are in the genotype. In the present section, a few philosophical aspects of this model are further developed. It is subdivided into three subsections: (1) aspects of manifestation; (2) gradual unfoldment and the stages of evolution; and (3) systems of interactions, vital field, and the implicate order.

6.1. *Aspects of manifestation*

In analysing the manifestation of evolution, we must distinguish various aspects which can be summarized under the heading 'self-expression'. Self-expression' is introduced here on the analogy of Portmann's 'Selbstdarstellung', a term that pertains to the (ontogenetic) transformation by which a hidden molecular order manifests itself as organic form (cf. Portmann, 1959: 422–425; see also: Kugler, 1967: 33–36). 'Self-expression' includes 'adaptation', by which term I mean the condition, of an organism (or a structure), of being adapted, of showing particular fitness for particular situations which are part of its environment. The term includes, however, several other aspects, of which the following are mentioned here.

a. In the course of evolution, organisms conform to the type of organization inherited, by which they are determined regardless of the environment (Grehan, 1984: 14). These types of organization probably include earlier adaptations which are no longer functional.

b. Evolution is also determined by certain fundamental properties of matter and life. Theories are, e.g., developed with reference to shapes organisms are likely or unlikely to possess (Thom, 1975; Thom, Lejeune & Duport, 1978; Saunders, 1984: 258–261). Early stages of evolution were certainly closely connected with the evolution of matter, such as the origin of ordered structures and the increase in complexity in physical systems (see, e.g., Fox, 1984; Lima-de-Faria, 1983; Matsuno, 1984; Prigogine, 1980; Prigogine & Stengers, 1984; Wicken, 1984).

c. As will be mentioned below, in subsection 6.3., certain characters of matter and life could represent fundamental characters of the manifestation of an unfolding implicate order.

6.2. *Gradual unfoldment and the stages of evolution*

The evolution of organic life is apparently characterized by the alternation of periods of rapid transformation and comparative stability, as a result of which a gradual (step by step) succession can be distinguished. Each of these levels represents the realization of potentialities, and includes new potentialities (as mentioned above, the evolutionary potentialities are in the genotype which is interpreted cybernetically as a program of action and reaction); the development of new potentialities is an important unsolved problem. In the case of the early stages of evolution, up to the origination of the regna and phyla, evolutionary mechanisms (the unfoldment of the potentialities and the modes of manifestation) must have been very different for each of the stages. Organic evolution started with the origination of primordial life, and progressed when living substances began to exist as definite bodies, and when reliable methods of self-replication and bisexual reproduction arose. The first important ascent in the scale of increasing complexity consisted in the origination of multicellular organisms, whilst the origin of many phyla is probably closely connected with the subsequent origination of segmentation. Starting from primitive segmented types, it is often not very difficult to interpret the subsequent evolution of a group as resulting from changes in systems of interactions (Van der Hammen, 1981a; see essay II). A concise survey of the early stages of evolution (research in this field is beyond my experience) is given below.

a. It is now generally accepted that primordial life must have had two properties: the capacity of harnessing energy, and the capacity of growth and reproduction. As mentioned above (in subsection 4.1.), these completely different properties cannot have been simultaneously present in a single homogeneous substance. The capacity of harnessing energy required the appearance of complex molecules functioning as catalysts. The capacity of reproduction and growth required the appearance of complex molecules capable of continuous self-perpetuation without diminution. The theory of the origin of life consequently presupposes the association or the associated development of the two completely different and highly complex types of molecules (see, e.g., Bonik, Grasshoff, Gutmann & Maier, 1978: 162–163; Fox, 1984; Good, 1981: 115–158; Lima-de-Faria, 1983; Matsuno, 1984; Oparin, 1953; Wicken, 1984). It may be remarked here that, according to Thom (1975: 286), the synthesis of life requires the execution of an infinite number of ordinary catastrophes, controlled by a well-established plan, before the stable metabolic situation can be established, and so an infinite number of local syntheses in a well-defined spatiotemporal arrangement.

b. Individuals in organic nature (see Bonik, Grasshoff, Gutmann & Maier, 1978: 163–165; J.S. Huxley, 1912; Jeuken, 1952; Koepcke, 1973: 81–116; Medawar, 1957) first arose when living substances began to exist as definite bodies. The single cell must have been the historical basis of organic individuality. To the properties of primordial life (the capacity of harnessing energy, and the capacity of growth and reproduction), the cell added the cell-wall (with double function: protection and absorption). In this way, the first organic individuals constituted unities with an increased independence, consisting of a certain diversity of inter-dependent parts (i.e. parts with integrated functions). Originally, organic individual and individual death were not inseparably connected. Cells divided, and two new individuals took the place of the parent individual. The life of cell-lineages, which arose in this way (and still arise in the case of extant Prokaryota and Protista), can be characterized as indeterminate. Death, in this case, could be accidental (i.e. arise from any cause whatsoever). Individual death, and the occurrence of a mortal body, apparently first arose in organisms like *Volvox*: colonies of cells, with a differentiation of the members, in which some cells are specialized and have an asexual reproductive function. In the act of sexual reproduction, the major part (the body) of the colony must die. In the course of evolution, natural individual death became increasingly associated with the age-specific decline of vital faculties.

c. The earliest living organisms certainly had a capacity for self-replication, although this process probably was inexact (the genotype could have been liable to lose identity, so that the progeny departed more or less widely from the condition of their parent). In the course of evolution, more reliable methods of replication arose, but reproduction in ancestral Prokaryotes probably was asexual (in this group, the counterpart of the eukaryote nucleus consists of a circular strand of DNA). Some mechanism of genetic recombination probably arose early in the prokaryotic line. In extant Bacteria, part of a strand of DNA of one individual can be transferred to another (in various ways), and recombination takes place by exchange of material. It can be assumed that the ancestral Eukaryote was a haploid asexual unicellular alga, with a nucleus, a set of linear chromosomes, and an efficient nuclear division resulting in an exact replication of the genome (mitosis). The origin of sex presupposes the fusion of two haploid cells and the attainment of true diploidy, the origination of nuclear divisions which halve the number of chromosome complements (meiosis), and the origination of gametes, followed by fertilization. During meiosis, genetic material can be exchanged between homologous chromosomes (recombination). The primary consequence of sex is the diversification of progeny (see Bell, 1982). In Protista, the somatic cell can develop into a macrogamete (two macrogametes subsequently fuse), or it can develop into a gametangium with several microgametes. In multicellular algae, gametes arise in special cells (gametangia); the gametes can be morphologically similar (isogamy) or present differences in size (anisogamy). In the case of anisogamy, the macrogamete can become a non-motile egg-cell which is fertilized by a free-swimming microgamete or spermatozoid (oogamy).

d. Starting from Protista, the first important ascent in the scale of increasing complexity of living nature consisted in the origination of multicellular organisms. This step could have been realized in various ways. Colonies of uniform cells could have arisen from Protista, followed by a division of labour among the cells; the colonies could subsequently have been transformed into true individuals. It is, however, also possible that division of labours set in at the beginning, and that no colony preceded the multicellular individual. According to Bonik, Grasshoff & Gutmann (1976: 132–136, fig. 3), the evolution of multicellular animal organisms must have started from multinucleate, ciliate Pro-

14

tista with internal jelly-formation; by division into compartments, the formation of uninucleate cells constituting an epithelium, and increased solidity (because of the internal jelly-formation), the earliest multicellular animal organisms are supposed to have arisen. Although evolution included also regression
and extinction, the further ascent generally consisted in an increase in differentiation, complexity, harmonious integration and efficiency, and an increase in
independence and individual autonomy. Progressive evolution could be restricted to a perfection of adaptation within the framework of a plan of construction, or pertain to a more important ascent to a higher (more complex)
plan of construction. In the last-mentioned case, various major plans of construction could succeed each other in time (see J.S. Huxley, 1912: 99–114;
Koepcke, 1973: 99–101; Overhage, 1957).

e. The origin of many phyla is probably closely connected with the origin of
segmentation. In the course of evolution, segmentation probably arose more
than once, and perhaps in different ways. It can pertain to one or more internal
systems; true segmentation includes segmentation of muscular system and coelom, pseudosegmentation is restricted to other systems. There are several theories pertaining to the origin of segmentation. According to the pseudosegmen-
tation-theory, true segmentation arose from pseudosegmentation (in groups
without coelom), and a coelom could subsequently have arisen from gonads.
According to the cyclomere-theory (which is associated with the enterocoel-
theory, pertaining to the origin of the coelom), segmentation arose from radial
symmetry; coeloms originated, in this case, from caeca (blind diverticula of the
alimentary canal), as in the embryonic development of some groups; and from
a number of radially symmetric caeca, a series of bilaterally symmetric coelom
compartments arose. According to the bilaterogastraea-theory (which is also
associated with the enterocoel-theory), bilaterally symmetric coelom compartments arose from caeca; this change was associated with the evolution of adult
benthic forms from pelagic forms. According to the chain-theory, segmentation originated from asexual reproduction; new individuals arose in the
posterior part of a parent individual, and as soon as these remained attached,
they became part of a single segmented individual. Bonik, Grasshoff & Gutmann (1976a), who regarded the coelom as a hydroskeleton, connected the origin of segmentation with undulating movements and the associated origination
of coelom compartments and muscular segmentation; the complete internal
anatomy must subsequently have been accommodated to this metameric
scheme. In my opinion, the origination and subsequent evolution of segmentation is an event of which a hypothetical explanation in terms of genetic control
is of paramount importance (cf. Elder, 1979, 1984); a similar explanation of
Lankester's so-called laws of metamerism (see Lankester, 1902, 1904) would
also be of great interest.

6.3. *Systems of interactions, vital field, and the implicate order*

As mentioned in subsection 5.3., Von Baer (1876) regarded evolution already as a process manifesting itself in the course of ontogeny, and resulting in important modifications. Modern authors increasingly recognize that the process of development can indeed affect the course of evolution (Saunders, 1984: 255), and Garstang (quoted by Løvtrup, 1984: 116) even wrote that ontogeny creates phylogeny. Generally, transmutations will be more important as they manifest themselves earlier in ontogeny.

The group of sciences which study the mechanisms through which transformation, in the course of ontogeny, takes place, is called epigenetics (cf. Waddington, 1957: 13–14, 130, 166, 189–190, Fig. 4). It is generally assumed that gene expression in the phenotype is a result of genetic and developmental interactions (cf. García-Bellido, 1983); in this connection, mention is sometimes made of regulatory genes which control the activities of structural genes. The epigenetic code, however, constitutes an area where formidable technical difficulties obstruct the biochemical approach; Kauffman (1973, 1987) and Elder (1979, 1984) recently suggested that abstract coding theory could be of value in the deduction of the control systems involved. In terms of epigenetics, evolution must be attributed to changes in systems of interactions. Many examples of numerical changes, in which changes in epigenetic coding are supposed to be involved, are mentioned in my paper on *Numerical Changes and Evolution in Actinotrichid Mites* (Van der Hammen, 1981a; see essay II); an interpretation of these data in terms of information theory could perhaps disclose some characters of the epigenetic and the underlying evolutionary code. An interesting example of the co-ordination of changes in a system of interactions with changes in the environment was published by Boelé & Van der Hammen (1982; a summary is now included in essay II); this example pertained to the variation, along the coast, from North to South, of the number of ungues in the claws of an Oribatid mite from European salt-marshes.

I have repeatedly pointed out the importance of segmentation for the understanding of the evolution of many phyla (Van der Hammen, 1981: 36; 1985: 402, 406; 1986: 21–31). In a study of the embryonic development of *Drosophila*, García-Bellido, Lawrence & Morata (1979) concluded that an animal can be made up of a number of different, but fundamentally 'homologous' (i.e. homonomous) units or 'compartments' (segments and subsegments) which represent variations of a basic genetic theme. A few genetic control elements can specify a large number of different compartments, and decisions are made very early in development (see also Lima-de-Faria, 1983: 969–971, Fig. 39.21). The concepts of homotype (homotypes are the homonomous parts of which an archetype can be composed) and metamorphosis (the transformation of these parts), well-known from studies by Goethe and Owen (see essay V), can now be explained and re-established in the light of these conclusions. A similar, although more complicated explanation can be given of the concept of meta-

16

morphosis in the sense of transformation during postembryonic development; in this case, each of the compartments of an instar has the potentiality to transform itself repeatedly (and sometimes radically) in the course of ontogenetic time (see essay III). Kauffman (1973, 1987) and Elder (1979, 1984) further developed the hypothesis that the logic of epigenetic control may be based on a binary combinatorial epigenetic code.

It is evident that evolution can now be understood as change in epigenetic coding; this implies that evolutionary control may be based on an evolutionary code which belongs to deeper layers of the hierarchic structure (apparently a very complicated hierarchy of relatively simple laws) underlying development and evolution. It is, probably, at the deepest level of this hierarchic structure that changes in the environment are co-ordinated with evolutionary changes (the ultimate nature of the co-ordination is really a metaphysical problem).

Development and evolutionary possibilities are apparently preformed in the molecular order, whilst this order gradually manifests itself epigenetically. For this reason, we can sometimes speak of the evolutionary program and the evolutionary potentialities of the genotype (see essay II, where it is demonstrated that there is also a probability of manifestation). In several of the following essays, these conclusions are further developed, particularly on the basis of data from chaetotaxy, the life-cycle and the appendages.

In his Structural Stability and Morphogenesis, Thom (1975) has tried to attack, in an exclusively geometrical way, the problem of morphogenesis (which, as we have seen, is closely related to the problem of evolution). According to him, the fundamental problem of biology is a topological problem (topology is the mathematical discipline which deals with the passage from the local to the global). In his view (Thom, 1975: 151–152), all living phenomena can be seen as manifestations of a geometric object, the vital field ('champ vital', translated in the English edition by 'life field'). The first task is the geometric description of this field, the determination of its formal properties and its laws of evolution, while the question of the ultimate nature of the field (whether it can be explained in terms of known fields of inert matter) is really a metaphysical one. Thom (1975: 158–159) rightly remarked that the attitude of the reductionist is metaphysical (he postulates a reduction that has never been experimentally established). Thom's method of attributing a formal geometrical structure to a living being, in order to explain its stability, may be thought of as a kind of geometrical vitalism; it provides a global structure (within a multidimensional space) controlling the local details. But this structure can in principle be explained solely by local determinisms, theoretically reducible to mechanisms of a physico-chemical type. It is not admissible to explain local phenomena by the global structure. In a similar way, the evolution of life and its environment can be regarded as manifestations (projections) of a global geometric structure (the ultimate nature of which is a metaphysical question). In the case of evolution, the local phenomena are more difficult to investigate. In several of the following essays, however, many local phenomena are described and analysed, whilst ontophylogenetic models are prepared of the underlying evolutionary mechanisms.

In the metaphysical context of the ultimate nature of evolution, it may be interesting to mention the work of David Bohm who, in his Wholeness and the Implicate Order (Bohm, 1980), developed a theory of quantum physics, that treats the totality of existence, including matter and consciousness, as an unbroken whole, a multidimensional reality. (According to Bohm, a theory is a way of looking at nature, which guides our perceptions. Bohm's views of the unconscious background of consciousness are very similar to those developed by C.G. Jung.) The implicate order is regarded as a process of enfoldment and unfoldment in a higher-dimensional space. The explicate order of manifestation is a special distinguished case of a set of implicate orders. The various particles of quantum mechanics (of which the laws are statistical and do not determine individual events uniquely and precisely) have to be taken literally as projections of a higher-dimensional reality, which cannot be accounted for in terms of any force of interaction between them. Time is a projection of a multidimensional reality into a sequence of moments. Evolution (i.e. unfoldment) cannot be properly understood without considering the immense multidimensional reality of which it is a projection. Bohm criticized the modern biologist's strong belief in the fragmentary atomistic approach to reality (and the neglect of the revolutionary developments in modern physics). According to him, the study of life and mind are just the fields in which formative cause, acting in undivided and unbroken flowing movement, is most evident to experience and observation.

According to Bohm, the information content of the implicate order is enfolded and carried in a form of movement (waves, e.g.), in a way similar to a hologram. (A hologram is an interference pattern of which each part contains the information of the whole, although somewhat less detailed.) An order of this kind (it reminds us of Thom's vital field) could underly structures such as the morphogenetic and the chromosome field. According to Goodwin (1984: 228) 'a field is a spatial domain in which every part has a state determined by the state of neighbouring parts so that the whole has a specific relational structure. Any disturbance to the field, such as removal, reordering, or addition of parts, results in a restoration of the normal relational order so that one whole spatial pattern is reconstituted.' According to Lima-de-Faria (1986), the chromosome field displays the same type of properties as the field in embryology. 'It forms a structure, the chromosome pattern, which has the capacity of restoring the whole following a disturbance.' In both cases, information on the whole is, apparently, present in the parts.

When organic evolution is indeed conceived as a projection of a multidimensional reality, this implies that the most important determinant event is not constituted by the manifestation of change in a single individual, but by a process of unfoldment simultaneously projected in all individuals concerned. Interesting metaphysical perspectives are now opened for the study of the con-

18

nection between evolutionary program and changing environment, and for the study of the evolutionary programs of co-evolving partners (in cases like symbiosis, parasitism, etc.); in all these cases, associated enfoldment of information results in the associated projection of evolutionary changes.

A deeper understanding of unfoldment could perhaps be gained by an integrated structuralist study of manifestation, including not only evolutionary biology, but also analytical psychology (structural elements of the unconscious become conscious as archetypal images) and other related disciplines. In this context, it may finally be remarked that, according to Goethe, polarity constitutes one of the basic properties of manifestation ('Erscheinung').

References

Aristotle, De Anima: On the soul. Parva naturalia. On breath. With an English translation by W.S. Hett. – The Loeb Classical Library, Cambridge, Mass. (Harvard University Press), London (William Heinemann Ltd.), 1975: xviii + 528 p.

Aristotle, De Partibus Animalium: Parts of animals. With an English translation by A.L. Peck. – The Loeb Classical Library, London (William Heinemann Ltd.), Cambridge, Mass.(Harvard University Press), 1961: vi + 556 p.

Aristotle, Historia Animalium: The works of Aristotle, translated into English under the editorship of J.A. Smith & W.D. Ross, vol. 4, Historia animalium, by D'Arcy W. Thompson. – Oxford, 1910: xvi + 482 p.

Aristotle, Metaphysica: The metaphysics, books I-IX. With an English translation by H. Tredennick. – The Loeb Classical Library, Cambridge, Mass. (Harvard University Press), London (William Heinemann Ltd.), 1980: xl + 473 p.

Aristotle, Physica: The physics, books I-IV. With an English translation by Ph. W. Wicksteed & F.M. Cornford. – The Loeb Classical Library, London (William Heinemann Ltd.), Cambridge, Mass. (Harvard University Press), 1963: xc + 427 p.

Baer, K.E. von, 1876. Reden gehalten in wissenschaftlichen Versammlungen und kleinere Aufsätze vermischten Inhalts, vol. 2. Studien aus dem Gebiete der Naturwissenschaften. – St. Petersburg: xxvi + 408 p., 22 figs.

Bell, G., 1982. The masterpiece of nature. The evolution and genetics of sexuality. – London & Canberra (Croom Helm): 635 p., 43 figs., 54 tab.

Berg, L.S., 1926. Nomogenesis or evolution determined by law. – London (Constable & Company Ltd.): 477 p.

Boelé, F.E. & L. van der Hammen, 1982. Variaton in the number of ungues in Ameronothrus schneideri (Oudemans), an Oribatid mite. – Zool. Meded. Leiden 56: 169–191, Fig. 1, Tab. 1–5.

Bohm, D., 1980. Wholeness and the implicate order. – London, Boston & Henley (Routledge & Kegan Paul); xvi + 224 p.

Bonik, K., M. Grasshoff & W.F. Gutmann, 1976. Die Evolution der Tierkonstruktionen. – Natur und Museum 106: 120–143, Figs. 1–3.

Bonik, K., M. Grasshoff & W.F. Gutmann, 1976a. Die Evolution der Tierkonstruktionen III. Vom Gallertoid zur Coelomhydrolik. – Natur und Museum 106: 178–188, Fig. 1.

Bonik, K., M. Grasshoff, W.F. Gutmann & W. Maier, 1978. Hydraulik als Grundlage der Morphologie aller tierischen Lebewesen. – Natur und Museum 108: 162–174, Figs. 1–9.

Broad, C.D., 1979. Leibniz. An introduction. (Edited by C. Lewy). – Cambridge (University Press): xii + 175 p. (second edition).

Buffon, G.-L. Leclerc de, 1749. Histoire naturelle, générale et particulière, avec la description du Cabinet du Roi, vol. 1. – Paris: vii + 612 p.

Buffon, G.-L. Leclerc de, 1749a. Histoire naturelle, générale et particulière, avec la description du Cabinet du Roi, vol. 2. – Paris: iv + 603 p.

Buffon, G.-L. Leclerc de & L.J.M. Daubenton, 1753. Histoire naturelle, générale et particulière, avec la description du Cabinet du Roi, vol. 4. – Paris: xvi + 544 p.

Claghorn, G.S., 1954. Aristotle's criticism of Plato's 'Timaeus'. – The Hague (Martinus Nijhoff): xii + 149 p.

Conrad-Martius, H., 1944. Der Selbstaufbau der Natur. Entelechien und Energien. – Hamburg (H. Goverts Verlag): 434 p.

Conrad-Martius, H., 1960. Die Geistseele des Menschen. – München (Kösel-Verlag): 86 p.

Darwin, C., 1872. The origin of species by means of natural selection, or the preservation of favoured races in the struggle for life. – London (sixth edition): xxii + 458 p.

Dobzhansky, Th., F.J. Ayala, G.L. Stebbins & J.W. Valentine, 1977. Evolution. – San Francisco (W.H. Freeman and Company): xiv + 572 p.

Driesch, H., 1928. Philosophie des Organischen. – Leipzig (fourth edition): xvi + 402 p.

Elder, D., 1979. An epigenetic code. – Differentiation 14: 119–122, Figs. 1–2, Tab. 1–3.

Elder, D., 1984. Theory of epigenetic coding. – Journ. Theor. Biol. 108: 327–332.

Fisher, R.A., 1930. The genetical theory of natural selection. – Oxford (University Press). [I have used the second edition, New York (Dover Publications, Inc.), 1958: 291 p., Pls. 1–3, Figs. 1–11, Tab. 1–12.]

Fox, S.W., 1984. Proteinoid experiments and evolutionary theory. – In: M.-W. Ho & P.T. Saunders, Beyond neo-Darwinism. An introduction to the new evolutionary paradigm (London, etc.: Academic Press): 15–60, Figs. 2.1–2.3, Tab. 2.1–2.5.

García-Bellido, A., 1983. Comparative anatomy of cuticular patterns in the genus Drosophila. – In: B.C. Goodwin, N. Holder & C.C. Wylie, Development and evolution (Cambridge University Press): 227–255, Figs. 1–12, Tab. 1.

García-Bellido, A., P.A. Lawrence & G. Morata, 1979. Compartments in animal development. – Sci. Am. 241(1): 90–98.

Glass, B., O. Temkin & W.L. Strauss Jr. (eds.), 1959. Forerunners of Darwin: 1745–1859. – Baltimore (The Johns Hopkins Press). [I have used the Johns Hopkins Paperbacks edition, 1968: xxii + 471 p., 5 figs.]

Good, R., 1981. The philosophy of evolution. – Stanbridge, Wintorne, Dorset (The Dovecote Press): 182 p.

Goodwin, B.C., 1984. A relational or field theory of reproduction and its evolutionary implications. – In: M.-W. Ho & P.T. Saunders, Beyond neo-Darwinism. An introduction to the new evolutionary paradigm (London, etc.: Academic Press): 219–241, Figs. 9.1–9.6.

Grehan, J.R., 1984. Evolution by law: Croizat's 'orthogeny' and Darwin's 'laws of growth'. – Tuatara 27: 14–19.

Häring, N., 1956. The Creation and Creator of the world according to Thierry of Chartres and Clarenbaldus of Arras. – Arch. Hist. Doctrin. Moyen-Age 30: 137–216.

Hammen, L. van der, 1978. The evolution of the chelicerate life-cycle. – Acta Biotheoretica 27: 44–60, Figs. 1–6. [Cf. essay III].

Hammen, L. van der, 1981. Type-concept, higher classification and evolution. – Acta Biotheoretica 30: 3–48, Figs. 1–7. [Cf. essay V].

Hammen, L. van der, 1981a. Numerical changes and evolution in Actinotrichid mites (Chelicerata). – Zool. Verh. Leiden 182: 1–47, Figs. 1–13. [Cf. essay II].

Hammen, L. van der, 1983. Unfoldment and manifestation: The natural philosophy of evolution. – Acta Biotheoretica 32: 179–193.

Hammen, L. van der, 1985. A structuralist approach in the study of evolution and classification. – Zool. Meded. Leiden 59: 391–409, Figs. 1–5. [Cf. essay VI].

Hammen, L. van der, 1986. On some aspects of parallel evolution in Chelicerata. – Acta Biotheoretica 35: 15–37, Figs. 1–9. [Cf. essay IV].

Harris, C.L., 1981. Evolution, genesis and revelations. With readings from Empedocles to Wilson. – Albany (State University of New York Press): xii + 339 p., Figs. 1–3, 8 tab.

Herder, J.G. von, 1784. Ideen zur Philosophie der Geschichte der Menschheit, vol. 1. – Leipzig. [I have used the fourth edition, edited by H. Leider, 1841.]

Huxley, J.S., 1912. The individual in the Animal Kingdom. – The Cambridge Manuals of Science and Literature (Cambridge University Press): xii + 167 p., frontispiece, Figs. 1–16.

Huxley, J.S., 1942. Evolution. The modern synthesis. – London (George Allen & Unwin Ltd.). [I have used the fourth impression, London, 1945: 645 p.].

Huxley, Th.H., 1893. Darwiniana. Essays [= Collected essays, vol. 2]. – London: xii + 475 p.

Jeuken, M., 1952. The concept 'individual' in biology. – Acta Biotheoretica 10: 57–86.

Kauffman, S.A., 1973. Control circuits for determination and transdetermination. – Science 181: 310–318, Figs. 1–7, Tab. 1–5.

Kauffman, S.A., 1987. Developmental logic and its evolution. – BioEssays 6(2): 82–87, Figs. 1–3.

Koepcke, H.-W., 1973. Die Lebensformen (Grundlagen zu einer universell gültigen biologischen Theorie), vol. 1. – Krefeld (Goecke & Evers): xvi + 789 p., Figs. 1–374.

Kugler, R., 1967. Philosophische Aspekten der Biologie Adolf Portmanns. – Basler Beiträge zur Philosophie und ihrer Geschichte, 2: viii + 200 p.

Lamarck, J.-B.-P.-A. de Monet de, 1809. Philosophie zoologique ou exposition des considérations relatives à l'histoire naturelle des animaux, à la diversité de leur organisation et des facultés qu'ils obtiennent; aux causes physiques qui maintiennent en eux la vie et donnent lieu aux mouvements qu'ils exécutent; enfin, à celles qui produisent les unes le sentiment, les autres l'intelligence de ceux qui en sont doués. – Paris. [I have used the following edition: Nouvelle édition, revue et précédée d'une introduction biographique par Charles Martius. – Paris, 1873; vol. 1: lxxxiv + 412 p.; vol. 2: 431 p.]

Lankester, E.R., 1902. Arthropoda. – Encyclopaedia Britannica (tenth edition) 25: 689–701, 11 figs.

Lankester, E.R., 1904. The structure and classification of the Arthropoda. – Quart. Journ. Micr. Sci. (N.S.) 47: 523–582, Pl. 42.

Leibniz, G.W.: New essays on human understanding. Translated and edited by P. Remnant & J. Bennett. – Cambridge (University Press), 1981: xcvi + 527 p.

Leibniz, G.W.: Philosophical papers and letters. Edited by L.E. Loemker. – Dordrecht (D. Reidel Publishing Company), 1969: xii + 736 p. (second edition).

Liepman, H.P., 1981. The six editions of the Origin of Species. A comparative study. – Acta Biotheoretica 30: 199–214.

Liepman, H.P., 1982. De zes edities van Darwin's 'Origin of Species'. – Vakblad voor Biologen 62: 435–437, Figs. 1–2.

Lima-de-Faria, A., 1983. Molecular evolution and organization of the chromosome. – Amsterdam, New York, Oxford (Elsevier). [I have used the paperback edition, Amsterdam, 1986: xliv + 1186 p., figs.].

Lima-de-Faria, A., 1986. The chromosome field theory confirmed by DNA cloning and hybridization. – Osaka Workshop on Structuralism in Biology, Summaries: 36–42. [Also in: Rivista di Biologia 80(2): 266–268].

Loemker, L.E., 1969. [See under Leibniz, G.W.].

Lovejoy, A.O., 1936. The Great Chain of Being. A study in the history of an Idea. – Cambridge, Mass. & London (Harvard University Press). [I have used the fourteenth printing, 1978: xii + 382 p.]

Løvtrup, S., 1984. Ontogeny and phylogeny. – In: M.-W. Ho & P.T. Saunders. Beyond neo-Darwinism. An introduction to the new evolutionary paradigm (London, etc.: Academic Press): 159–190, Figs. 7.1–7.10.

Lyell, C., 1872. The principles of geology or the modern changes of the earth and its inhabitants considered as illustrative of geology. – London (eleventh and entirely revised edition); vol. 1: xx + 671 p.; vol. 2: xx + 652 p.

Matsuno, K., 1984. Open systems and the origin of protoreproductive units. – In: M.-W. Ho & P.T. Saunders, Beyond neo-Darwinism. An introduction to the new evolutionary paradigm (London, etc.: Academic Press): 61–88, Fig. 3.1.

Medawar, P.B., 1957. The uniqueness of the individual. – London (Methuen & Co. Ltd.): 191 p.

Meulen, J. van der, 1966. A Logos Creator at Chartres and its copy. – Journ. Warburg and Courtauld Inst. 29: 82–100, Pls. 31–32.

Oparin, A.I., 1953. The origin of life. – New York (Dover Publications, Inc.): xxviii + 270 p. [First published in Russian in 1936; first English translation published in 1938.]

Overhage, P., 1957. Der 'biologische Aufstief' und seine Kriterien. – Acta Biotheoretica 12: 81–114.

Pasteur, L., 1861. Mémoire sur les corpuscules organisés qui existent dans l'atmosphère. Examen de la doctrine des générations spontanées. – Ann. Sci. Natur., Zool. (4) 16: 5–98, Pl. 1. [Also published in: Ann. Chim. Phys. (3) 64 (1862): 5–110, Pls. 1–2. Œuvres de Pasteur, vol. 2 (Paris, 1922): 210–294, Figs. 1–31.]

Plato, Timaeus: Platon, Œuvres complètes, vol. 10, Timée, Critias. Texte établi et traduit par A. Rivaud. – Paris (Les Belles Lettres), 1963: 227 p.

Portmann, A., 1959. Zur Philosophie des Lebendigen. – In: F. Heinemann (ed.), Die Philosophie im XX. Jahrhundert. Eine enzyklopädische Darstellung ihrer Geschichte, Disziplinen und Aufgaben (Darmstadt: Wissenschaftliche Buchgesellschaft): 410–440. [I have used the third edition, Darmstadt, 1970.]

Prigogine, I., 1980. From being to becoming. Time and complexity in physical sciences. – San Francisco (W.H. Freeman and Company): xix + 272 p.

Prigogine, I. & I. Stengers, 1984. Order out of chaos. [I have used the Dutch translation, Amsterdam (Bert Bakker), 1985: 352 p.]

Remnant, P. & J. Bennett, 1981. [See under: Leibniz, G.W.]

Robinet, J.-B., 1761–1766. De la nature, vols. 1–4. – Amsterdam; vol. 1 (1761), xx + 456 p. (published anonymously); vol. 2 (1763), xvi + 444 p.; vol. 3 (1766), lvi + 288 p.; vol. 4 (1766), iv + 284 p.

Robinet, J.-B., 1768. Vue philosophique de la gradation naturelle des formes de l'être, ou les essais de la nature qui apprend à faire l'homme. – Amsterdam: iv + 264 p.

Saunders, P.T., 1984. Development and evolution. – In: M.-W. Ho & P.T. Saunders, Beyond neo-Darwinism. An introduction to the new evolutionary paradigm (London, etc.: Academic Press): 243–263, Figs. 10.1–10.3.

Stölzle, R., 1897. Karl Ernst von Baer und seine Weltanschauung. – Regensburg: xii + 688 p.

Teichmüller, G., 1877. Darwinismus und Philosophie. – Dorpat: vii + 90 p.

Thom, R., 1975. Structural stability and morphogenesis. – Reading, Mass. (Benjamin): xxviii + 348 p., figs., pls. [English translation, by D.H. Fowler, of: Stabilité structurelle et morphogénèse, Reading, Mass., 1972].

Thom, R., C. Lejeune & J.-P. Dupont, 1978. Morphogenèse et imaginaire. – Circé, Cahiers de Recherche sur l'Imaginaire, 8–9: 144 p., figs.

Vries, J. de, 1980. Grundbegriffe der Scholastik. – Darmstadt (Wissenschaftliche Buchgesellschaft): xii + 120 p.

Waddington, C.H., 1957. The strategy of the genes. A discussion of some aspects of theoretical biology. – London (Allen & Unwin): ix + 262 p., 37 + X figs.

Waesberghe, H. van, 1981. Naar een derde evolutie-model. – Vakblad voor Biologen 61: 50–56, Fig. 1, Tab. 1–2.

Waesberghe, H. van, 1982. Towards an alternative evolution model. – Acta Biotheoretica 31: 3–28, Figs. 1–4, Tab. 1–4.

Wicken, J.J., 1984. On the increase in complexity in evolution. – In: M.-W. Ho & P.T. Saunders, Beyond neo-Darwinism. An introduction to the new evolutionary paradigm (London, etc.: Academic Press): 89–112.

II. NUMERICAL CHANGES IN DEVELOPMENT AND EVOLUTION IN ACTINOTRICHID MITES (CHELICERATA)

1. Introduction

In the first essay, it was mentioned that evolution must be regarded as a co-ordination of internal and external factors (genotype and environment), as the realization of genotypic potentialities, and as the result of changes in systems of interactions manifesting themselves in the course of ontogeny. The second essay is entirely devoted to the last-mentioned evolutionary changes. It is restricted to numerical changes in patterns of setae and setiform organs and in segmentation, in Actinotrichid mites (a group of Chelicerata), and to the way these changes manifest themselves in the course of postembryonic development. The essay is based on a paper under nearly the same title, which I published about six years ago (Van der Hammen, 1981a); the text is, however, entirely revised and partly rewritten, in accordance with the development of my views and the place of the essay in the context of this book. The results of the essay could perhaps lead to an information processing approach in the study of development and evolution.

Numerous studies have been published on numerical changes in patterns of setae and setiform organs and in segmentation, in Actinotrichid mites. The extensive literature on the subject is highly specialized and scattered over a great number of papers; as a consequence of this, it has remained practically unobserved (except by a small circle of initiates). In the present essay, a compendious survey of the changes is given, with special attention to suppression and multiplication, and to the ontogenetic and evolutionary aspects of these changes. It may be remarked here that Actinotrichida (one of the two large groups of mites) constitute a chelicerate superorder which, according to my present views (Van der Hammen, 1977, 1982, 1986) is related to Palpigradi; these two groups together constitute the Epimerata, a chelicerate subclass. Anactinotrichida (the second large group of mites; it is not included in this study) are related to Ricinulei; these two groups together constitute the Cryptognomae, another chelicerate subclass (cf. Van der Hammen, 1977, 1979a, 1986).

Actinotrichida (cf. Fig. 2.1 and Van der Hammen, 1982) are Chelicerata of which the prosoma and the opisthosoma are broadly joined, whilst the mouthparts constitute a movable gnathosoma. The prosoma includes an unknown

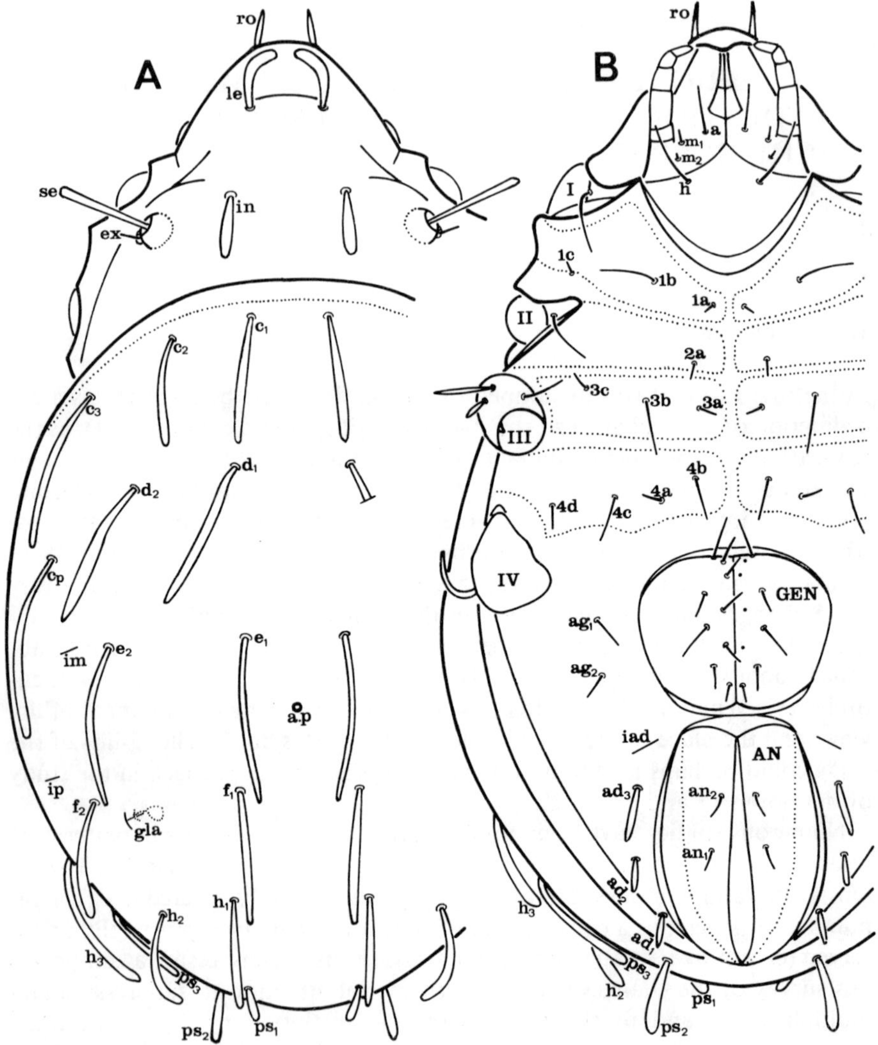

Fig. 2.1. Hermannia gibba (C.L. Koch), adult male; A: dorsal view; B: ventral view (legs, with the exception of the trochanter, removed); A, B: × 161.

number of precheliceral metameres and the segments I–IV; part of this segmentation is still distinctly recognizable, as epimera, at the ventral surface (segments III–VI, the segments bearing legs I–IV). The six pairs of appendages (chelicera, palp, four pairs of legs) are homonomous structures which belong to segments I–VI; metameric parallelisms are particularly distinct in the case of the legs. The opisthosoma is ancestrally composed of the segments VII–XVII, but segments XIII–XVII are suppressed during embryonic development and can reappear, with the exception of segment XVII, in the course of postembryonic development (hysteromorphosis). Each of these segments has its own base level: XIII is larval, XIV protonymphal, XV deutonymphal, and XVI tritonymphal; segment XVII, which adult base level, is not known but could still be present in unknown primitive species. Segment XVI is known from a few primitive groups only; in most Oribatid mites the last segment (including the anal opening) is constituted by segment XV, in other groups of Actinotrichid mites various numbers of segments are added, whilst in Tarsonemida the last segment is constituted by segment XIII. The opisthosoma can present homologues of five pairs of opisthosomatic appendages: the endites of the appendages of segments VIII–IX, constituting the eugenital lips (which border the eugenital opening); the exites of these appendages, which can constitute the progenital covers (they close the progenital chamber, which includes the eugenital opening); and often three pairs of genital papillae, which represent homologues of the opisthosomatic appendages of segments X–XII, and which can be included in the progenital chamber. The dorsal surface of the prosoma presents, ancestrally, six pairs of setae, of which the segmental origin is uncertain. The epimera present numbers of setae which can be of taxonomic importance. Chaetotactic patterns of the opisthosoma are also of taxonomic importance; the segmental origin of the opisthosomatic setae is not always known with certainty.

The greater part of the phenomena dealt with here have been investigated by the French acarologist F. Grandjean, in a period extending over some thirty-five years (Grandjean, 1938a, 1939, 1939a, 1939c, 1939d, 1939e, 1941, 1941a, 1942, 1942b, 1942c, 1943, 1943a, 1943b, 1946a, 1947a, 1947b, 1948a, 1948b, 1949, 1949a, 1949b, 1951, 1951a, 1952, 1954a, 1957, 1958, 1961b, 1961c, 1963, 1964, 1965, 1965a, 1971, 1971a, 1973, 1974).[1] Some of Grandjean's nearest collaborators continued his research, and published several partial summaries (Van der Hammen, 1962, 1963, 1964, 1974, 1975, 1978, 1979, 1980 1981a; Lions, 1964; Travé, 1973, 1974, 1977, 1979, 1979a; Coineau, 1973, 1974; Boelé & Van der Hammen, 1982). An early summary was published by Knülle (1957).

The present essay constitutes an introduction to the investigations in this field; it is the first general introduction dealing with all aspects of the subject, and the first general theory. The changes are associated here with changes in systems of genotypic interactions. Detailed definitions of the extensive terminology introduced by Grandjean and his collaborators are given in the general part of the *Glossary of Acarological Terminology* (Van der Hammen, 1980); in this glossary, numerous detailed references to the literature are also given. In the present essay numerical variations in clones, in populations, and in

---

[1] Grandjean's papers are collected and reprinted in my edition of his Complete Acarological Works (The Hague, Lochem, 1972–1976). This edition (in seven volumes) was published with introductions and a detailed index.

supraspecific taxa are dealt with first. After that, the ontophylogenetic aspects of numerical changes, and the phenomenon of priority are discussed. The evolution of chaetotactic patterns, and the ontophylogenetic aspects of genital chaetotaxy and of solenidiotaxy are discussed in separate sections. Thereupon, the numerical changes in a large taxonomic group (the Oribatid superfamily Nothroidea) are dealt with. Conclusions, general theoretical considerations, and suggestions for further research are given in the final section.

The problems dealt with here are partly related to genetics. For this reason the following data on Actinotrichid reproduction and cytogenetics are mentioned. Actinotrichid reproduction can be sexual or asexual. Parthenogenesis is a common phenomenon, known from all four orders. It is either arrhenotokous, thelytokous or deuterotokous. The last-mentioned type is rare; it is known from the family Listrophoridae (Acaridida). In cases of thelytoky, several mechanisms are known for re-establishing the diploid chromosome number. Karyotypes of many Actinotrichid species (among them spider mites) are known. The chromosomes of several species of Actinotrichida are known or supposed to have a diffuse centromere (kinetochore) (Pijnacker & Ferwerda, 1972).

Data on regulatory mechanisms (systems of interactions) are summarized by Dobzhansky, Ayala, Stebbins & Valentine (1977). Regulatory genes control the activity of other (structural) genes. Little is known about the mechanisms that regulate gene activity in higher organisms, although Britten and Davidson have proposed an interesting model that is consistent with the requirements of the genetic system and with our present knowledge of cell biology (cf. Dobzhansky, Ayala, Stebbins & Valentine, 1977: 29, 256–258, Fig. 8.11). Changes in systems of interactions are associated with evolutionary phenomena such as the shifting of changes in character states from one part of an ontogeny to another (Dobzhansky, Ayala, Stebbins & Valentine, 1977: 206). García-Bellido (1983) studied the developmental genetic mechanisms underlying variations in cuticular patterns (among them patterns of setae) in *Drosophila* (Diptera). Kauffman (1973, 1987) studied control circuits for determination and transdetermination in *Drosophila*, and the properties of regulatory networks in eukaryotic development and the implications for development and evolution. Elder (1979, 1984) studied the genetic control of the development of segmentation in *Hirudo* (Annelida); he introduced the hypothesis that it is based on a binary epigenetic code.

2. Numerical variations in clones

Individual variations in clones of *Platynothrus peltifer* (C.L. Koch) have been studied by Grandjean (1948, 1948a, 1948b, 1950, 1954a, 1958, 1971a, 1973, 1974). He had previously discovered the occurrence, in wild populations of this species, of discontinuous variations pertaining to numerous organs of body and appendages.

Platynothrus peltifer is an Oribatid mite (classified with the superfamily Nothroidea) reproducing by thelytokous parthenogenesis. It is excellently suited for the study of individual variations, because it can easily be reared in the laboratory, whilst it can also easily be studied under the microscope (it is not too small, its movements are slow, and its legs are robust). Grandjean (1948: 450—457) described his methods of rearing and the foundation of clones. Grandjean (1948a: 1—4) published thereupon a survey of the individual variations observed in a clone; this was followed by theoretical considerations on the nature of the variations (Grandjean, 1948b: 1—4). He described also the behaviour of the species, based on the observation of specimens in cultures (Grandjean, 1950: 229—231). Ontophylogenetic aspects of his observations were discussed in a theoretical paper (Grandjean, 1954a: 427—429). A study of the so-called accessory setae of the tarsus (those setae which are of postlarval origin in the case of legs I—III, and of postprotonymphal origin in the case of leg IV),[2] based on clones of *Platynothrus peltifer*, was subsequently published by him (Grandjean, 1958: 280—301, Figs. 1—6). A profound analysis of his observations on clones of the species (observations dating from the period 1947—1958) was carried out in the last years of his life. Three parts of this analysis have been published (Grandjean, 1971a, 1973, 1974), whilst the manuscript of a fourth part (not projected as the final paper of the series) was not yet completed at his death. The series, although unfinished, constitutes an extraordinary contribution to science, in which Grandjean had attained the borders of evolutionary morphology.

In the present section, a summary is given of some aspects of Grandjean's studies (followed by a new interpretation). Other aspects, especially those with reference to ontophylogeny, will be dealt with below. Important parts of Grandjean's results (especially those pertaining to the arrangement of setae) are not discussed here, because they do not refer to the subject of the present essay.

Grandjean studied the variation of a great number of organs, such as setae, solenidia and lyrifissures, in adults as well as immature forms (larva, protonymph, deutonymph and tritonymph). The organs included in his study were all idionymous, i.e. capable of receiving a designation which is not collective, and which enables recognition among other homonomous organs (either in the course of postembryonic development, or in a comparative study of a group of related species). Among the variations, two types of deviations from the normal condition could be distinguished: anomalies (exceptional variations without evolutionary significance) and so-called vertitions. A vertition is a discontinuous individual variation (absence or presence) pertaining to an idionymous organ normally present, resp. absent in specimens of the species, at the same level of postembryonic development, provided that this variation has an

---

[2] Leg IV is suppressed in the larva; it appears in the protonymph.

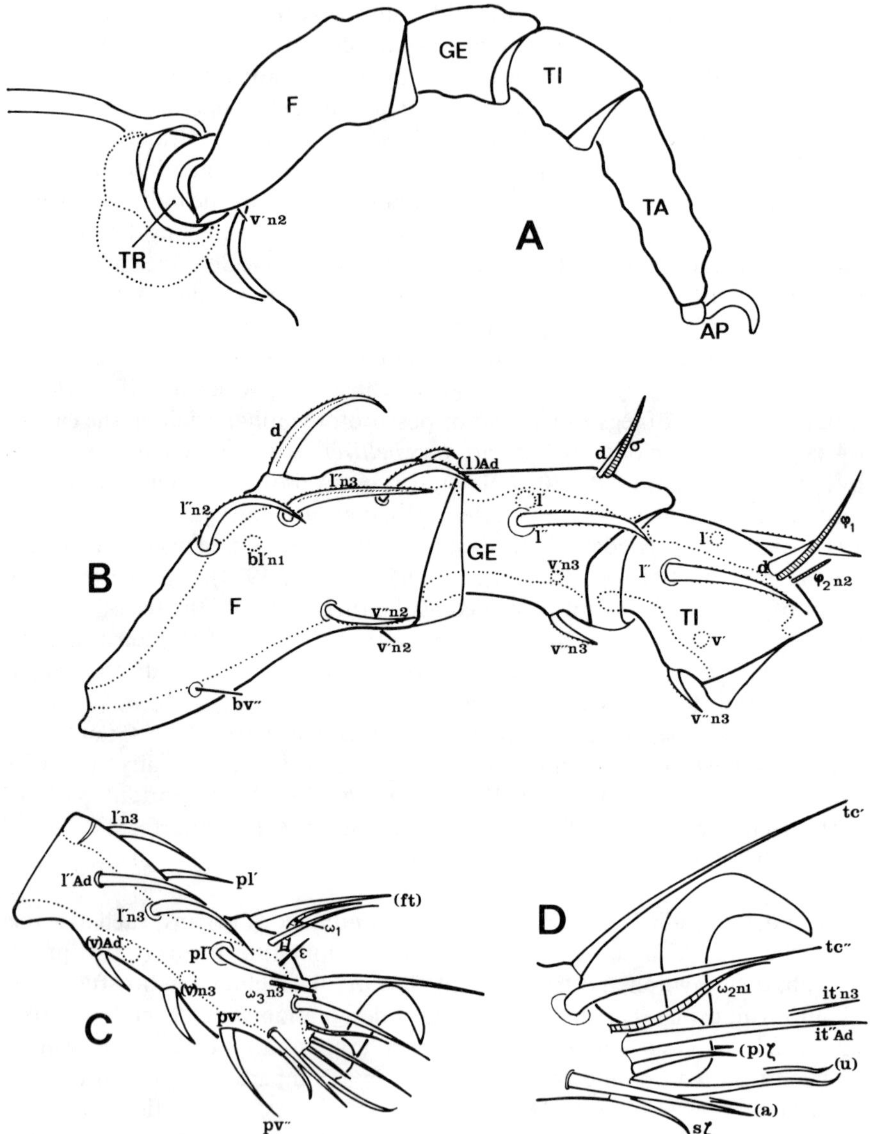

Fig. 2.2. Lateral (antiaxial) view of right leg I of the adult of *Platynothrus peltifer* (C.L. Koch); A: leg and acetabulum; B: femur, genu and tibia; C: tarsus and apotele; D: terminal part of tarsus, and apotele; A: × 232; B, C: × 368; D: × 687. Notations include a reference to the base level of the phaneres (n1 = protonymphal; n2 = deutonymphal; n3 = tritonymphal; Ad = adult; phaneres not accompanied by an ontogenetic notation are of larval origin). Vertitions pertaining to the larval phaneres are extremely rare; frequency of vertitions increases progressively in the case of protonymphal, deutonymphal, tritonymphal and adult phaneres. In the case of *Platynothrus peltifer*, vertitions pertaining to the lateral (*l*) and ventral (*v*) postlarval (= accessory) setae of tarsus and femur, are especially numerous.

evolutionary significance and is fundamentally unilateral (although there is a possibility of symmetrical manifestation). Vertitional changes are not immediately hereditary (in contradistinction to mutations); the offspring inherits a certain evolutionary potentiality and a probability of manifestation. The probability of existence can be calculated statistically; it can be expressed in a value varying from 1 (absolute presence) to 0 (absolute absence). Evidently, no vertitions are found in the cases of absolute presence or absence. Vertitions generally manifest themselves bilaterally only in cases of a high probability of existence. In a clone of *Platynothrus peltifer*, in which 50.000 single organs were investigated, the total percentage of vertitions appeared to be 1.3. The greater part of the vertitions (cf. Fig. 2.2) pertained to organs of postlarval origin (not yet present in the larva, but first appearing in protonymph, deutonymph, tritonymph or adult).

The most important characters of a vertition are constituted by the fundamentally unilateral manifestation, the heredity in the offspring of a probability of manifestation, and the evolutionary significance. Evidently, vertitions do not pertain to permanent changes in the genome (they are not identical with mutations), but refer to recurrent changes in an operator/repressor system (a system of interactions) of which the control system is still unknown. Some of the properties of the system, viz., the presence of an evolutionary potentiality and a probability of manifestation, could perhaps give an indication of the nature of this evolutionary code.

3. Comparative studies of vertitions in various populations of one species

The ancestral claw of Actinotrichid mites (cf. Grandjean, 1939e: 539–546) was tridactyl in all stases,[3] as still found in Endeostigmata and Palaeosomata (*Aphelacarus acarinus* (Berlese)). In most Oribatid mites, the tridactyl claw has, however, been subject to regression, and has become monodactyl in all immature stases; in the adults the claws can be mono-, bi- or tridactyl. In the course of evolution, monodactyly can arise by suppression of the two lateral ungues (*ol'* and *ol''*), homobidactyly by the suppression of the central unguis (*oc*), heterobidactyly by suppression of one of the lateral ungues (*ol'* or *ol''*).

Several cases of numerical regression of the claws have been investigated in detail. Grandjean (1939e: 542–543; 1961c: 539–555; 1965: 106–110) studied the regression of the ungues of the claws in *Nothrus silvestris* Nicolet, an Oribatid mite of which the reproduction is parthenogenetic. He based his study on material from various parts of France, Switzerland and Italy. In the genus *Nothrus* (Fig. 2.3), the lateral ungues can be subject to numerical regression.

---

[3] A stase is an idionymous instar. A complete definition can be found in the Glossary of Acarological Terminology (Van der Hammen, 1980).

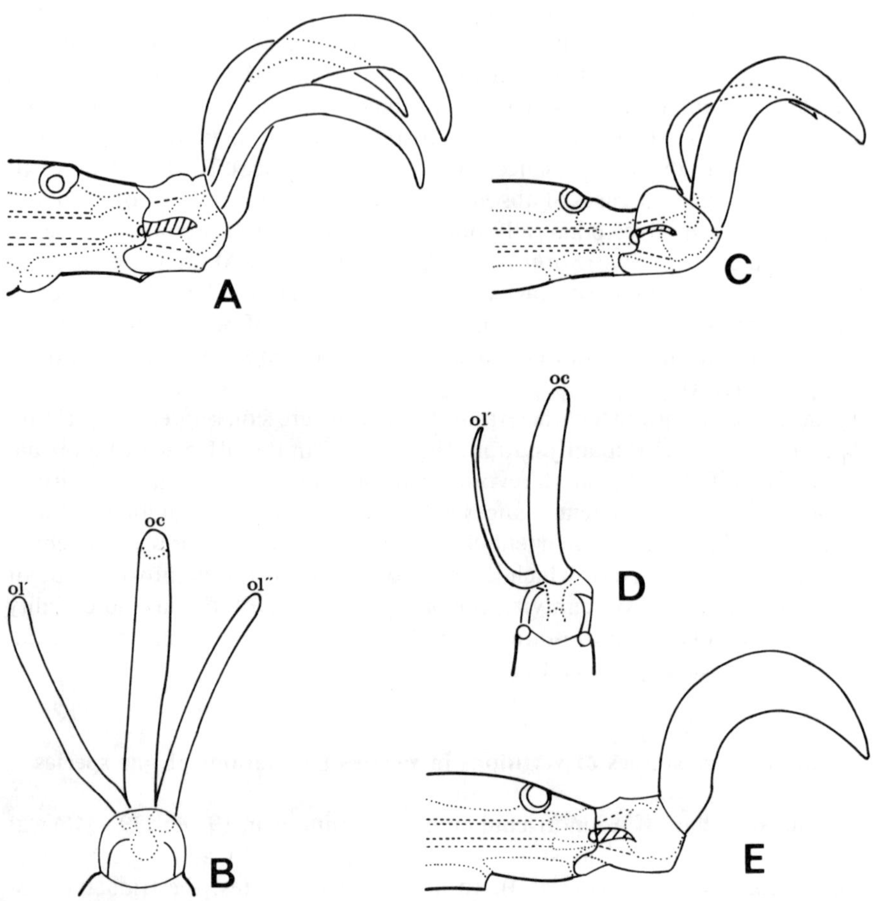

Fig. 2.3. Terminal part of right leg I of the adult in three *Nothrus* species; A, B: *Nothrus palustris* C.L. Koch; C, D: *Nothrus silvestris* Nicolet; E: *Nothrus pratensis* Sellnick; A, C, E: lateral (antiaxial) view; B, D: dorsal view; A–E: × 687.

Adults of *Nothrus silvestris* are generally bidactyl (the ungues mostly found are *ol'* and *oc*; *ol'* is stronger than *ol"*); specimens with monodactyl and tridactyl claws are, however, also found (monodactyl claws are regularly found; the occurrence of tridactyl claws is rare, and appears to be restricted, in a specimen, to one leg at the utmost). The evolution of the ungues is vertitional: regression manifests itself partly asymmetrically; in many cases different legs present different claws; and the specimens of one population seldom belong to one claw-type. Bidactyly appeared to be statistically predominant in wooded areas, monodactyly in open country. Statistically, the lateral ungues of legs I and II disappear before those of legs III and IV.

According to Grandjean (1965: 109), unguis *ol"* is stronger than *ol'* in the case of *Nothrus anauniensis* Canestrini & Fanzago (*ol'* is stronger in the case of *Nothrus silvestris*).[4] In *Nothrus anauniensis*, specimens generally present tridactyl claws; bidactyl claws are, however, also found, whilst specimens with monodactyl claws are extremely rare.

Lions (1964: 41−65) studied the variations of the number of ungues in *Rhysotritia ardua* (C.L. Koch), an Oribatid mite. The reproduction of this species is also parthenogenetic (cf. Lions, 1967: 282). Lions based his investigations on material from three different biotopes in the surroundings of Aix-en-Provence, in France. *Rhysotritia ardua* is predominantly bidactyl at leg I and tridactyl at legs II−IV. Regression has, consequently, started at leg I. In bidactyl claws of this species, the ungues are always *ol''* and *oc*. Monodactyl claws are also found. Claw-types are, generally, distributed over the legs in a more or less asymmetrical and apparently irregular way; evidently the variations must be attributed to vertitions. Homonychy (the occurrence of one claw-type in all legs of a specimen) appeared to be more common in specimens from the open country, heteronychy (the occurrence of various claw-types in the legs of one specimen) appeared to be more common in shaded country. The order of regression, in the legs, presented differences of a very complex nature.

Matsakis (1967: 590−631) made a mathematical analysis of the data collected by Lions. He discovered the existence of distinct, although very complicated connections between the variations; a number of elementary processes appeared to have acted together and in superposition. Statistically, *ol''* appeared to be stronger than *ol'*; there was, however, also a double priority because the lateral ungues of legs II and III were stronger than those of legs I and IV. Matsakis discovered also the presence of a certain balance (association of opposite and mutually compensatory variations) between *ol'* and *ol''* in neighbouring symmetrical or adjacent legs. Evidently, we are dealing here with an extremely complicated and balanced manifestation of changes in a system of interactions.

Boelé & Van der Hammen (1982) investigated the variation in the number of ungues in *Ameronothrus schneideri* (Oudemans), an Oribatid mite from salt-marshes, with bisexual reproduction. In the genus *Ameronothrus*, tridactyl claws are predominantly found in species living on hard substrates,

---

[4] A case in which the same unguis disappears in all legs pertains to evolutionary conformity (Grandjean, 1961: 210−217; cf. also Grandjean, 1939b: 40−41): the regression affects homonomous organs with corresponding metameric position. In Actinotrichida legs I and II are generally directed to the front, legs III and IV backward (a derived character state; ancestrally, the legs were orientated perpendicularly to the axis of the body). As a result of this, a number of phaneres at the paraxial side of the legs (the anterior side in the case of legs I and II, the posterior side in the case of legs III and IV), tend to disappear. In the case of evolutionary conformity, evolution operates as if the legs still had the ancestral position; it must be assumed to be produced internally, without external cause. In the case of evolutionary non-conformity, when an organ at the anterior side of legs I and II, and the homonomous organ at the posterior side of legs III and IV have disappeared, it must be assumed that the evolutionary changes are connected (moreover) with external causes.

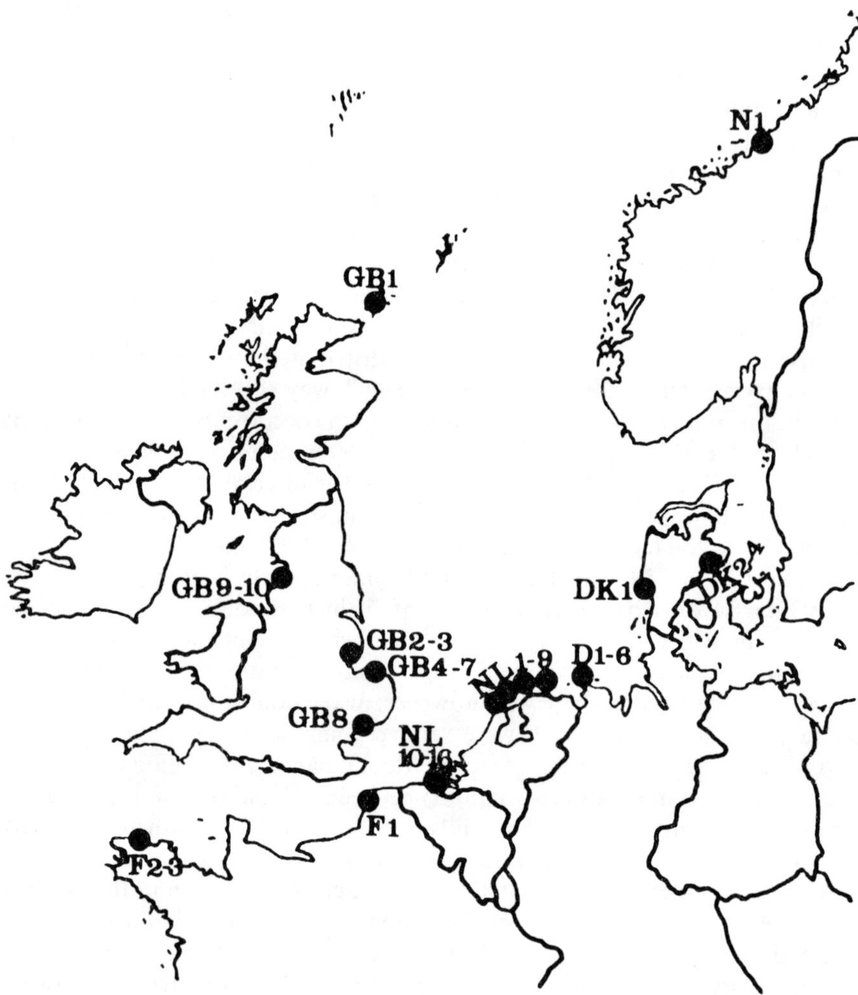

Fig. 2.4. Sketch-map of N.W. Europe, showing the localities (with station-numbers) in which material of *Ameronothrus schneideri* (Oudemans) was collected.

monodactyl claws in species living on soft substrate, although there are some notable exceptions. In *Ameronothrus schneideri*, the number of ungues is variable, and the variations are vertitional. The study by Boelé & Van der Hammen was based on specimens from Norwegian, Danish, German, French and English coasts (Fig. 2.4). Among the total number of specimens investigated, only 2.6 per cent presented tridactyl claws at all legs, whilst 45.2 per cent presented monodactyl claws at all legs. No specimens were found with bidactyl claws type *B'* (*oc* and *ol'*) or type *B''* (*oc* and *ol''*) on all legs. It appeared that the claws of the two anterior pairs of legs were more often monodactyl than the claws of the two posterior pairs of legs; the claws of the two posterior pairs of legs appeared to be more often tridactyl than the claws of the two anterior pairs of legs. Statistically, *ol'* was slightly stronger than *ol''*. There appeared to be a very interesting co-ordination between changes in the environment and the changes in the system of interactions, underlying the vertitional changes in claw-types: there was a very distinct increase, in the number of completely monodactyl specimens, from north to south, as far as the Slack estuary (south of Calais) in France. This tendency was apparently not continued further south; in material from Roscoff (Brittany), in France, the number of completely monodactyl specimens was again very small. The presence of *ol'* (1.0 = absolute presence; 0 = absolute absence) gradually decreased from Norway to N. France, from 0.71 to 0.01, and suddenly increased in the material from Brittany (0.86); the values for *ol''* were respectively 0.61, 0.01 and 0.77. Similar values were found in the material from Great Britain. It has not yet been analysed which environmental factors were associated with the changes.

The results of section 3 confirm and extend the conclusions of section 2. It is now also evident that the manifestation of vertitional changes can be co-ordinated with changes in the environment, and that, in series or patterns of symmetrical and homonomous elements, the changes are also influenced by distributional information.

4. Vertitional evolution in supraspecific taxa

The full evolutionary significance of vertitional variations can only be understood when a comparative study is made of numerous related species. Many data are, e.g., known in the case of the regressive and vertitional evolution of the so-called accessory setae of the tarsus in the Oribatid superfamily Nothroidea.

In Oribatid mites, the tarsal setae (with the exception of the iteral setae) can be subdivided in two groups: (1) fundamental setae (fastigial, tectal, proral, ungual, subungual, antelateral, primiventral and primilateral setae), of which the base level is generally larval in the case of legs I–III, and protonymphal in the case of leg IV; and (2) accessory setae (lateral and ventral setae) which appear in the course of the postembryonic development (after the larval stase, in the case of legs I–III; after the protonymphal stase, in the case of leg IV). Accessory setae are inserted proximally of the fundamental setae; they are arranged in verticils and files. In the adult, the total number of accessory setae of all four legs together varies from 104 to 0 (cf. Grandjean, 1984b: 3–4). The highest number is known from *Heminothrus targionii* (Berlese) (Fig. 2.5 A, B), whilst the total disappearance is, for instance, known from the genera *Malaconothrus, Trimalaconothrus* (Fig. 2.5 C) and *Trhypochthoniellus* (all taxa mentioned here belong to the superfamily Nothroidea). The accessory

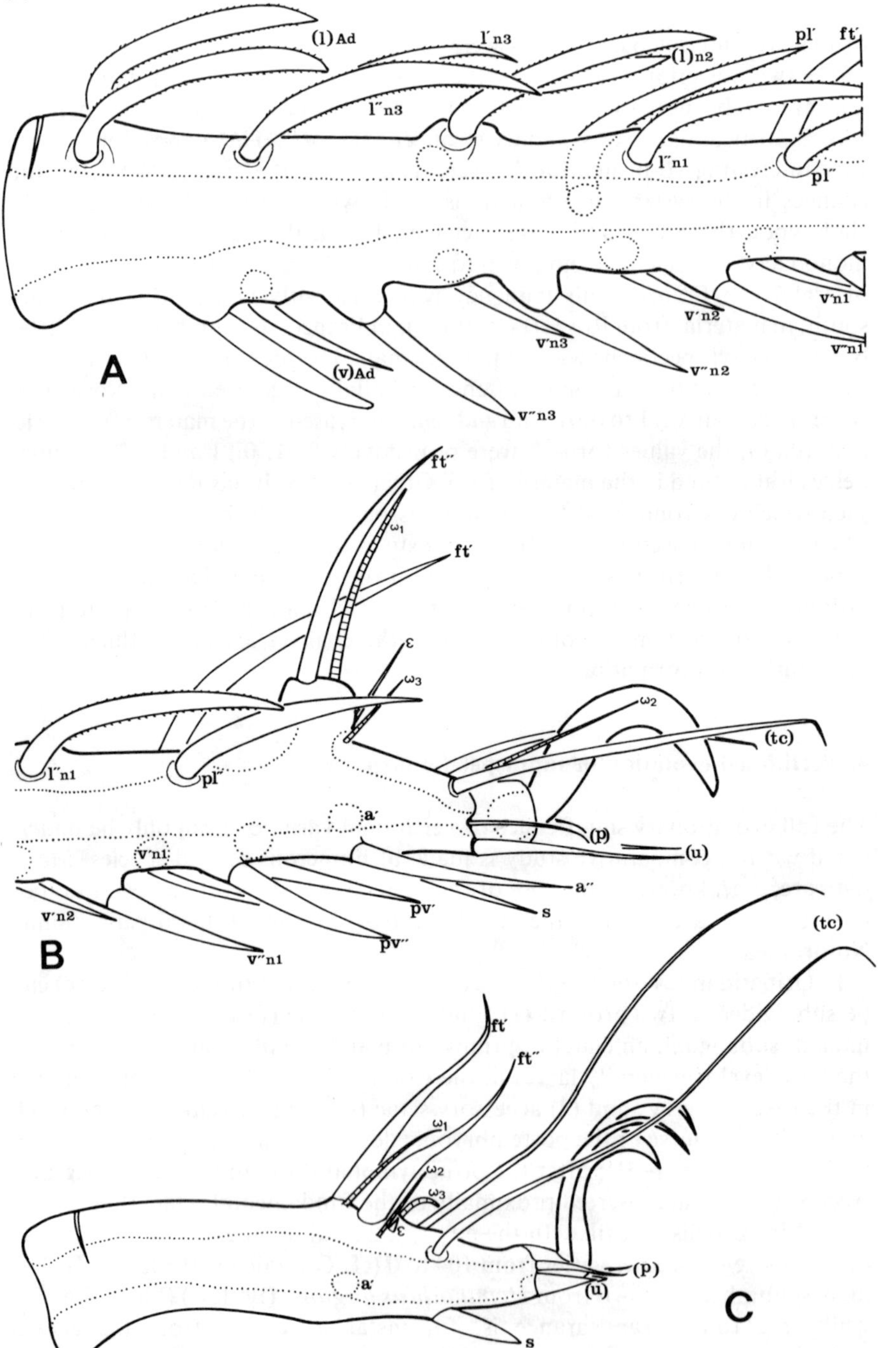

(l)Ad
l´n3
l´´n3
(l)n2
pl´
ft´
l´´n1
pl´´
v´n1
v´n2
v´´n1
v´n3
(v)Ad
v´´n2
v´´n3
A
ft´´
ω₁
ft´
ε
ω₃
ω₂
(tc)
l´´n1
pl´´
a´
(p)
(u)
a´´
v´n1
s
v´n2
pv´
v´´n1
pv´´
B
ft´
ft´´
(tc)
ω₁
ω₂
ω₃
ε
a´
(p)
(u)
s
C

setae are suppressed as part of a regressive and vertitional evolution. Its course can be studied, in a comparative way, in the group as a whole, because various stages of the regression and many vertitions are known.

A second example of a vertitional evolution, also from the superfamily Nothroidea, and pertaining to the femur of leg IV, was mentioned by Grandjean (1939a: 3). Femur IV of several Nothroid species, among which *Hermannia reticulata* (Thorell) and *Nothrus palustris* C.L. Koch, presents three setae. The median one of these setae (its base level is deutonymphal in the first-mentioned, tritonymphal in the last-mentioned species) has disappeared in many other species of the group. It is, generally, also absent in *Trhypochthonius tectorum* (Berlese). At one locality, however, Grandjean collected, among other specimens, two adults in which this seta was still present. The occurrence, in the two specimens, constitutes a very rare vertition.

Evidently, supraspecific taxa are characterized by evolutionary potentialities, underlying systems of interactions, which in different species manifest themselves in a greater or less degree.

5. Ontophylogenetic aspects of numerical changes

The relations between ontogeny and phylogeny in Actinotrichid mites have been studied by Grandjean (1942c, 1946a, 1947a, 1947b, 1951, 1951a, 1954a, 1957)[5] and Van der Hammen (1962, 1963, 1964, 1978). Before that time many morphologists and taxonomists adhered (and even adhere until now) the theory of recapitulation.

Numerical changes often manifest themselves in the course of postembryonic ontogeny. In a comparative study of the development of various species of a natural group, homologous changes may be found to take place either at various levels, or always at the same level of development.[6]

The changes, manifesting themselves in the course of postembryonic devel-

---

[5] Grandjean probably developed his first ideas with reference to the relation between ontogeny and phylogeny during his critical reading of De Beer's Embryology and Evolution (published in 1930; the evidence was presented again in De Beer's Embryos and Ancestors; cf. De Beer, 1958). Grandjean owned a copy of the translation into French by Rostand, in the margin of which he had made numerous critical notes. Grandjean's final views are strongly different from those of De Beer. Not until Grandjean's detailed analysis did the study of ontogeny approach the field of evolutionary cytogenetics.

[6] The postembryonic development of Actinotrichida is ancestrally characterized by the occurrence of six levels (stases): prelarva, larva, protonymph, deutonymph, tritonymph and adult; the prelarva has, in all known cases, been subject to regression (cf. Van der Hammen, 1978).

←

Fig. 2.5. Lateral (antiaxial) view of tarsus and apotele of right leg I of the adult in two species of Nothroidea; A, B: *Heminothrus targionii* (Berlese); A: proximal part of tarsus; B: distal part of tarsus, and apotele; C: *Trimalaconothrus grandis* Van der Hammen; A–C: × 687.

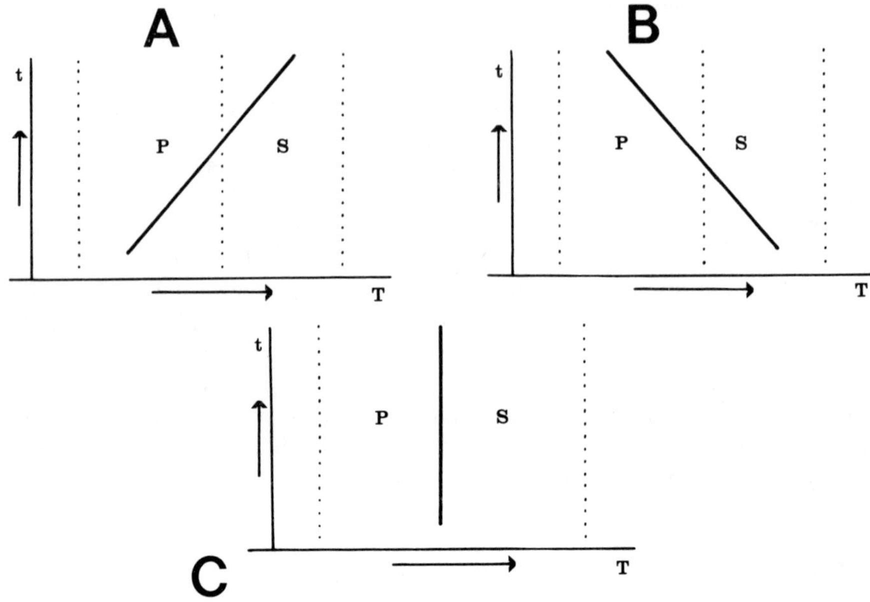

Fig. 2.6. Ontophylogenetic diagrams of an evolution *PS* (the change from the ancestral character state *P* to the derived character state *S*). The phylogenetic time *T* is on the horizontal axis, the ontogenetic time *t* on the vertical axis. In the diagram the ancestral character state *P* is separated from the derived character state *S* by a heavy line (the line of chronological separation). This line can be ascendant (A), descendant (B) or vertical (C). The dotted vertical lines represent ontogenies. In A three ontogenies are represented, of which the left presents the ancestral character state *P* only, the right the derived character state *S* only; the second from the left cuts the line of chronological separation: the ontogeny first presents the derived character state, then the ancestral character state (and is consequently the reverse of phylogeny). In B three ontogenies are also represented: one presenting *P* only, one presenting *S* only, and one (cutting the line of chronological separation) presenting first *P* and then *S*. In C all ontogenies present either *P* or *S*; none of them cuts the line of chronological separation.

opment, can refer to the replacement of an ancestral character state by a derived character state (ontogeny and phylogeny having the same direction), or to the replacement of a derived character state by an ancestral character state (the directions of ontogeny and phylogeny being opposed); the change can also refer to the replacement of a youthful character state by an adult character state (the last-mentioned type is not discussed in the present section).

Ontogenies can be plotted, side by side, in a diagram, in an attempt to prepare a model of the evolution of a character. In such an ontophylogenetic diagram, the ontogenetic time (*t*) is on the vertical axis, the phylogenetic time (*T*) on the horizontal axis. The ontogenies are arranged in such a way that, in the diagram, the ancestral character states are found at the left, the derived character states at the right. In simple cases, three types of ontophylogenetic diagrams can be distinguished (Fig. 2.6).

(1) The line separating, in the diagram, the ancestral character states from the derived character states (the line of chronological separation) presents, from left to right, a descendant course. An ontogeny, cutting this line, first presents the ancestral character state, then the derived character state (Fig. 2.6 B).

(2) The line of chronological separation presents, from left to right, an ascendant course. An ontogeny, cutting this line, first presents the derived character state, then the ancestral character state (ontogeny being the reverse of phylogeny) (Fig. 2.6 A).

(3) The line of chronological separation is vertical, and is cut by no ontogeny; the evolutionary change is always restricted to one level (when it does not take place at the level in question, it does not take place at all) (Fig. 2.6 C).

These three types will be explained, discussed and illustrated with the help of some examples (Figs. 2.7, 2.8).

(1a) In Actinotrichid mites, the dorsal seta d of genu and tibia is often associated with a solenidion. This association, evolutionarily considered, is only temporary because, as a result of the association, the seta tends to disappear. There are species in which d is absent in the adult, other species in which this seta is absent in the adult and one or more nymphs, and finally those in which it is absent in all stases. The disappearance of d must be regarded as a regression, i.e. a derived character state. The regression starts with the adult, and consequently belongs to the descendant type (Fig. 2.7 G, H).

(1b) In various species of the Oribatid genera *Hydrozetes* and *Limnozetes* the sensillus is absent (or vestigial), and the bothridium regressive, at the level of the adult. In the immature stases of these species, and in all stases of other, related species, sensillus and bothridium are normally developed. The regression constitutes a derived character state; it has started with the adult, and consequently belongs to the descendant type (Fig. 2.7 E, F). This type of trichobothridial regression is generally referred to as the *Hydrozetes* type.

(2a) In many species of Oribatid mites, seta v' of genu I appears in the course of postembryonic ontogeny. It can, according to Grandjean (1942c), appear in the protonymph (such as in *Eulohmannia ribagai* (Berlese), *Heminothrus targionii* (Berlese), *Nothrus palustris* C.L. Koch and *Hermannia reticulata* (Thorell)), in the deutonymph (such as in *Parhypochthonius aphidinus* Berlese, *Nanhermannia nanus* (Nicolet), *Camisia segnis* (Hermann), *Poroliodes farinosus* (C.L. Koch) and *Ceratoppia bipilis* (Hermann)), in the tritonymph (such as in *Eniochthonius minutissimus* (Berlese), *Trhypochthonius tectorum* (Berlese), *Fuscozetes fuscipes* (C.L. Koch) and *Pelops acromios* (Hermann)), or in the adult (such as in *Rhysotritia ardua* (C.L. Koch), *Trimalaconothrus* spec. and *Scheloribates laevigatus* (C.L. Koch)). In other species of Oribatid mites (such as in *Hypochthonius rufulus* C.L. Koch, *Meristacarus porcula* Grandjean, and *Trhypochthoniellus setosus* Willmann) it has completely disappeared. Late appearance as well as complete disappearance constitute a regression, i.e. a derived character state. Regression starts with the protonymph and

38

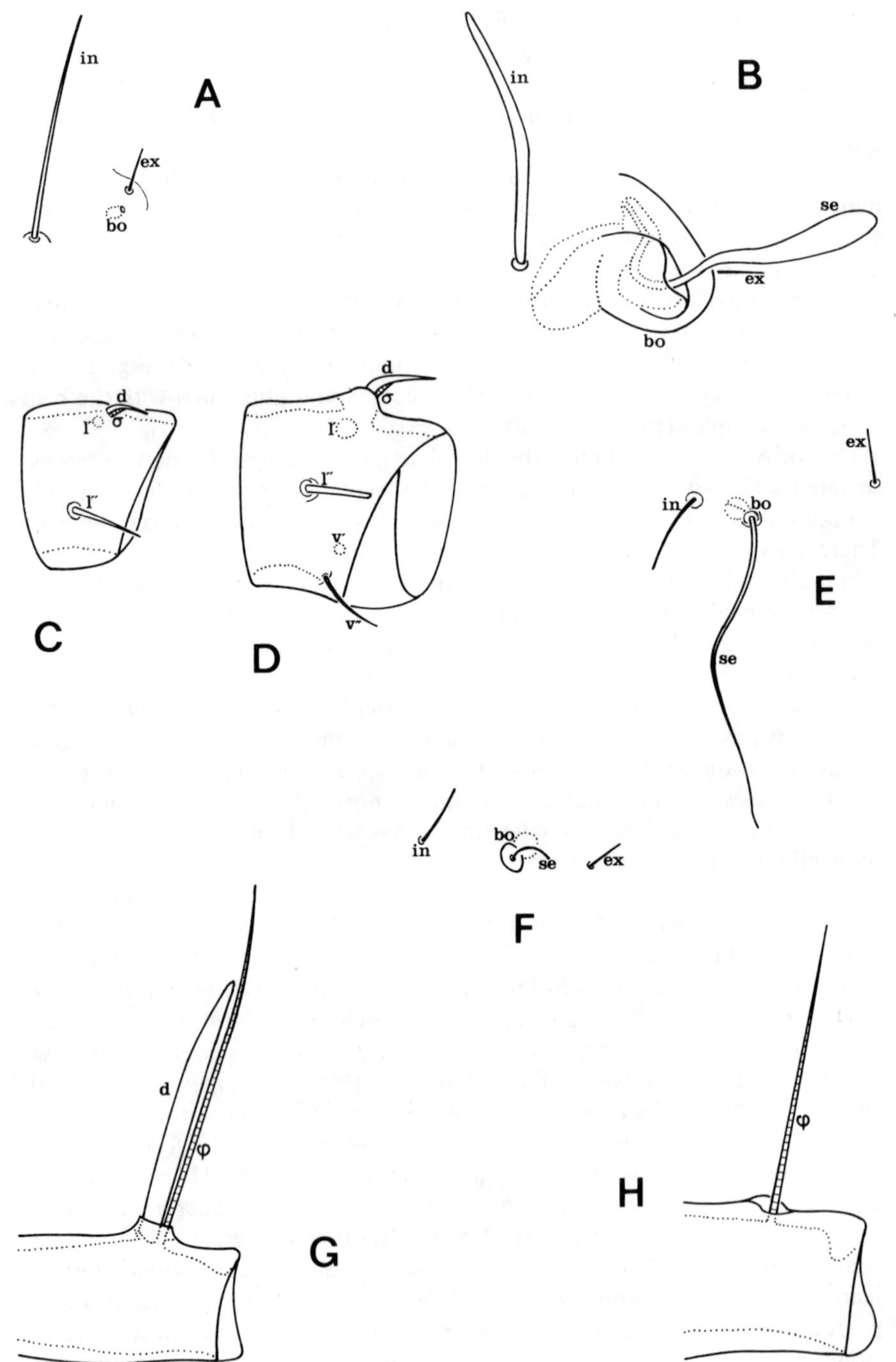

in
ex
bo
A
in
se
ex
bo
B
d
r
r
C
d
r
r
v
v
D
ex
in
bo
se
E
in
bo
se
ex
F
d
φ
G
φ
H

gradually affects all stases by extension of the ontogenetic retardation, finally resulting in complete suppression (Fig. 2.7 C, D).

(2b) In many species of Oribatid mites the sensillus, and often also the bothridium, have been subject to regression in a way different from the *Hydrozetes* type. The regressive evolution can pertain to the larva; to larva and protonymph; to larva, protonymph and deutonymph; to larva, protonymph, deutonymph and tritonymph; and even to all stases. The regression (regressive shape or complete absence) represents the derived character state. In postembryonic ontogeny the derived character state manifests itself before the ancestral character state (the ontogeny of this character being the reverse of phylogeny). In an ontophylogenetic diagram the line of chronological separation is ascendant. This type of trichobothridial regression is called the *Camisia* type (Fig. 2.7 A, B).

(3) In Actinotrichid mites, the so-called paraproctal segments appear either in the course of postembryonic development at fixed levels, or they do not appear at all. The pseudanal segment (segment XIII), if present, is of larval origin; the adanal segment (segment XIV), if present, is of protonymphal origin; the anal segment (segment XV), if present, is of deutonymphal origin; and the peranal segment (segment XVI), if present, is of tritonymphal origin. Consequently, the terminal segment of the adult, which surrounds the anus (or the uropore), can be constituted by: segment XIII (Tarsonemida, Tydeidae, Raphignathoidea, Tetranychoidea; and probably Cheyletoidea, Labidostommidae), segment XIV (Caeculidae, Anystidae, *Torpacarus*, Acaridida), segment XV (many Endeostigmata, most Oribatid mites), segment XVI (some Endeostigmata, some primitive Oribatid mites). The postembryonic development of the paraproctal region of *Archegozetes magna* (Sellnick), an Oribatid mite, is represented in Fig. 2.8; the figure shows the base level of the segments XIII, XIV and XV. For each of these segments ontophylogenetic diagrams (with the two character states: absence and presence) can be plotted, in which the line of chronological separation is vertical (and is not cut by any of the ontogenies).

In many cases, ontophylogenetic diagrams are much more complicated. The following three types of complicated diagrams are mentioned here:

(1) In disharmonic evolutions pertaining to regressive stases (elattostases, calyptostases) both states (ancestral as well as derived) can be absent at the level of the regressive stase; this absence is called carency. The case is illustrated in

---

←

Fig. 2.7. Changes in character states, in the course of postembryonic ontogeny, in Oribatid mites. A, B: *Platynothrus peltifer* (C.L. Koch), interlamellar seta, trichobothrium (or bothridium only) and exobothridial seta of the right side, viewed obliquely from above; A: tritonymph; B: adult. C, D: *Trhypochthonius tectorum* (Berlese), lateral (antiaxial = posterior) view of the genu of right leg I; C: deutonymph; D: tritonymph. E, F: *Hydrozetes lacustris* (Michael), interlamellar seta, trichobothrium and exobothridial seta of the right side, viewed obliquely from above; E: tritonymph; F: adult. G, H: *Damaeus onustus* C.L. Koch, lateral (paraxial = posterior) view of the terminal part of right tibia IV; G: tritonymph; H: adult. A, B, G, H: × 368; C–F: × 687.

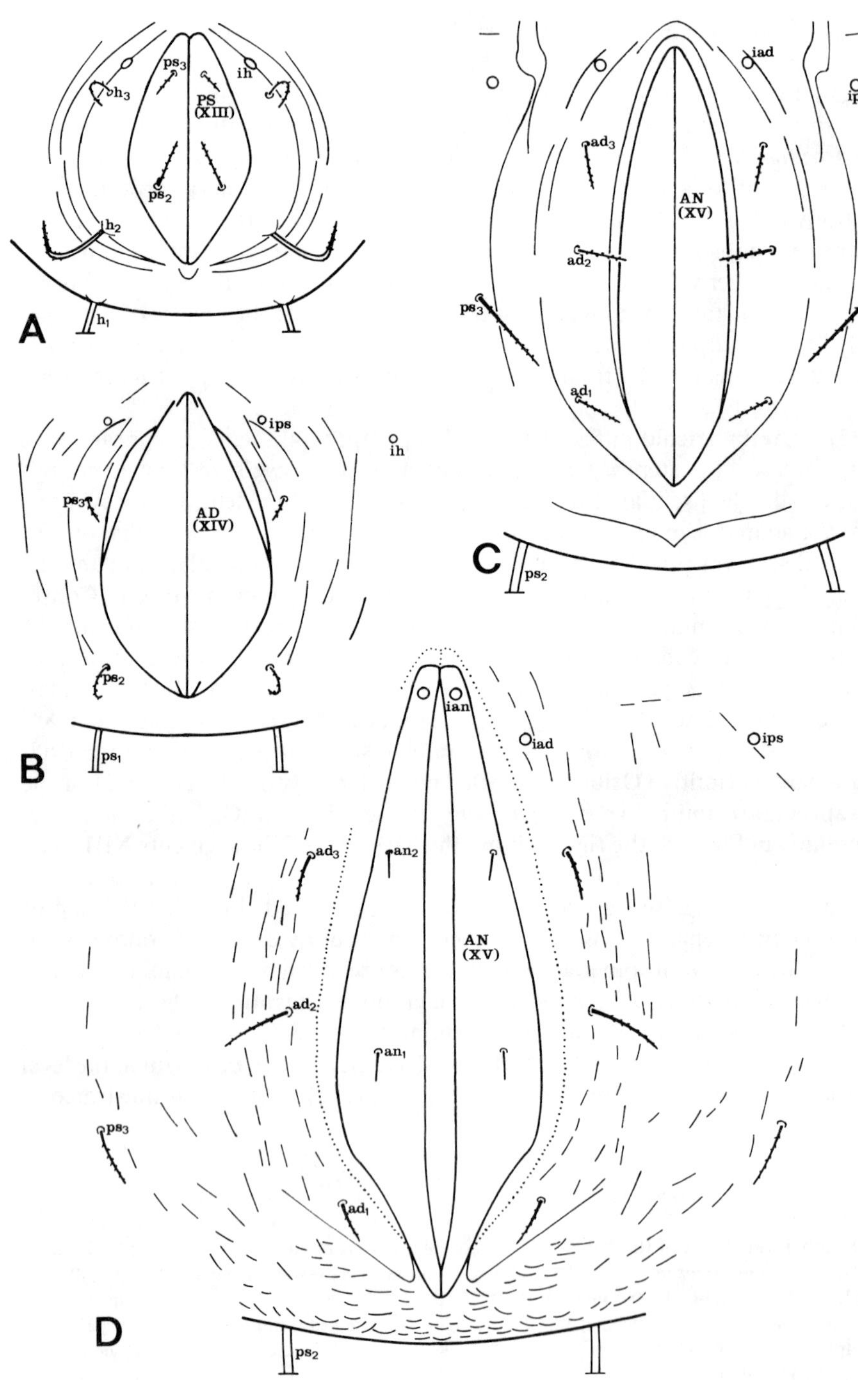
ps3
ih
h3
PS
(XIII)
ps2
h2
A
h1
iad
ips
ad3
AN
(XV)
ad2
ps3
ad1
ps2
C
ips
ih
ps3
AD
(XIV)
ps2
B
ps1
ian
iad
ips
ad3
an2
AN
(XV)
ad2
an1
ps3
ad1
D
ps2

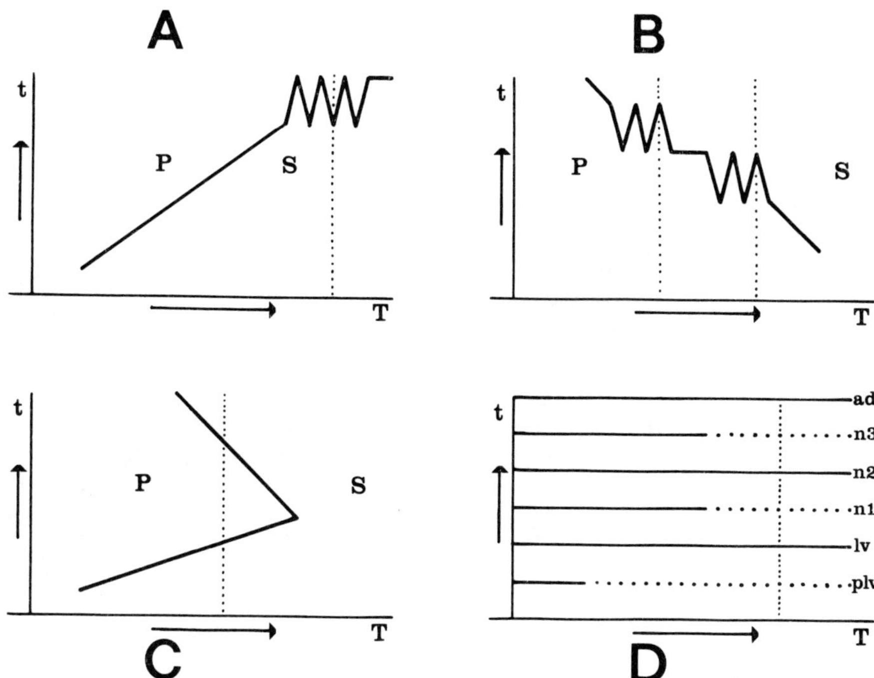

Fig. 2.9. Ontophylogenetic diagrams, A: ascendent evolution with vertitional period; the evolution represented here pertains to unguis *ol ″* of leg I in Nothroidea (cf. Fig. 2.3), and the dotted vertical line represents the postembryonic ontogeny of *Nothrus silvestris* Nicolet (the vertitional period is represented by a zig zag line). B: descendant evolution with two vertitional periods; the evolution represented here pertains to seta *d* of tibia IV in the genus *Xenillus*, and the two dotted vertical lines represent postembryonic developments of species in which this seta can be vertitionally present in respectively deutonymph and tritonymph (the vertitional periods are represented by a zig zag line). C: disharmonic evolution caused by the action of two evolutionary mechanisms; the line of chronological separation presents an ascendant and a descendant part; the evolution represented here pertains to seta *d* of tibia IV in the family Damaeidae; the dotted vertical line represents the postembryonic development of *Damaeus clavipes* (Hermann) (cf. Fig. 2.10). D: evolution of the postembryonic ontogeny in Trombidei, characterized by the development of regressive stases at the levels of prelarva, protonymph and tritonymph (represented by horizontal dotted lines); the vertical dotted line represents the postembryonic ontogeny of *Balaustium florale* Grandjean and many other species of this group; at the levels of the regressive stases, both states (ancestral as well as derived) of many characters are simultaneously absent.

←

Fig. 2.8. Postembryonic development of the paraproctal region of *Archegozetes magna* (Sellnick); A: larva; B: protonymph; C: deutonymph; D: tritonymph. In the larva, the paraproctal segment is constituted by the pseudanal segment (*PS = XIII*); in the protonymph it is constituted by the adanal segment (*AD = XIV*); in the deutonymph (and also in tritonymph and adult) it is constituted by the anal segment (*AN = XV*). Two segments (*XIV* and *XV*) are glabrous at the base level (i.e. the stase in which a character state makes its first appearance): there is atrichosy at two levels. The present addition of segments, in the course of postembryonic development, constitutes a special type of epimorphosis (hysteromorphosis); it does not pertain to anamorphosis. A–D: × 472.

Fig. 2.9 D, the ontophylogenetic diagram of *Balaustium florale* Grandjean; carency is found here at the levels of the calyptostasic proto- and tritonymphs.

(2) In a vertitional evolution (with one or more unstable periods), the line of chronological separation can be partly represented by one or more zig zag lines (a zig zag line representing a vertitional period). Vertitional periods can be found in ascendant as well as descendant evolutions (Fig. 2.9 A, B); generally, diagrams are simplified and the common unstable periods are not represented.

(3) In evolutions simultaneously influenced by two evolutionary forces, the ontophylogenetic diagram is generally also disharmonic, because the line of chronological separation can present an ascendant as well as a descendant part. An example of this type of disharmonic evolution is constituted by the evolution of seta *d* of tibia IV in the Oribatid family Damaeidae. The ontophylogenetic diagram is represented in Fig. 2.9 C; the postembryonic ontogeny, for the case of *Damaeus clavipes* (Hermann), is represented in Fig. 2.10. Seta *d* of tibia IV is absent in the protonymph as a result of the so-called protonymphal denudation of leg IV (in Actinotrichida, leg IV is absent in the larva, whilst the number of phaneres is strongly reduced in the protonymph). Seta *d* can, however, also disappear as a result of the association with a solenidion. The regression is ascendant in the first case, descendant in the second case. This results in a line of chronological separation with an ascendant and a descendant part; this line can be cut two times by an ontogeny.

It is now evident that changes in systems of interactions can manifest themselves in various ways in the course of postembryonic development and that different changes can even affect one and the same element.

6. Priority among members of a group of homonomous elements

In a taxon the resistance of the elements of a group to suppression or regression can present differences resulting in the numerical predominance of one element over another. This numerical predominance can be expressed in a priority list on which the elements are arranged in the order of decreasing strength (cf. Grandjean, 1938a: 1–5; 1941a: 1–5; 1942: 1–4; 1942a: 48–51; 1943a: 135–139; 1946a: 310–316; 1947: 79, 112; 1958: 307–308; 1961b: 624–626; 1964: 549–551). There are various approaches to the study of the relative strength and weakness of a group of elements.

The easiest approach is constituted by the comparison of species of a natural group: the most common element being at the same time the element with the greatest strength. In this case, priority can be either absolute (an organ never exists when a stronger organ, preceding it in a priority list, is absent), or statistical (an organ is, on the average, absent when an organ of greater strength is absent).

The second approach is ontogenetic: the strongest element of a group is

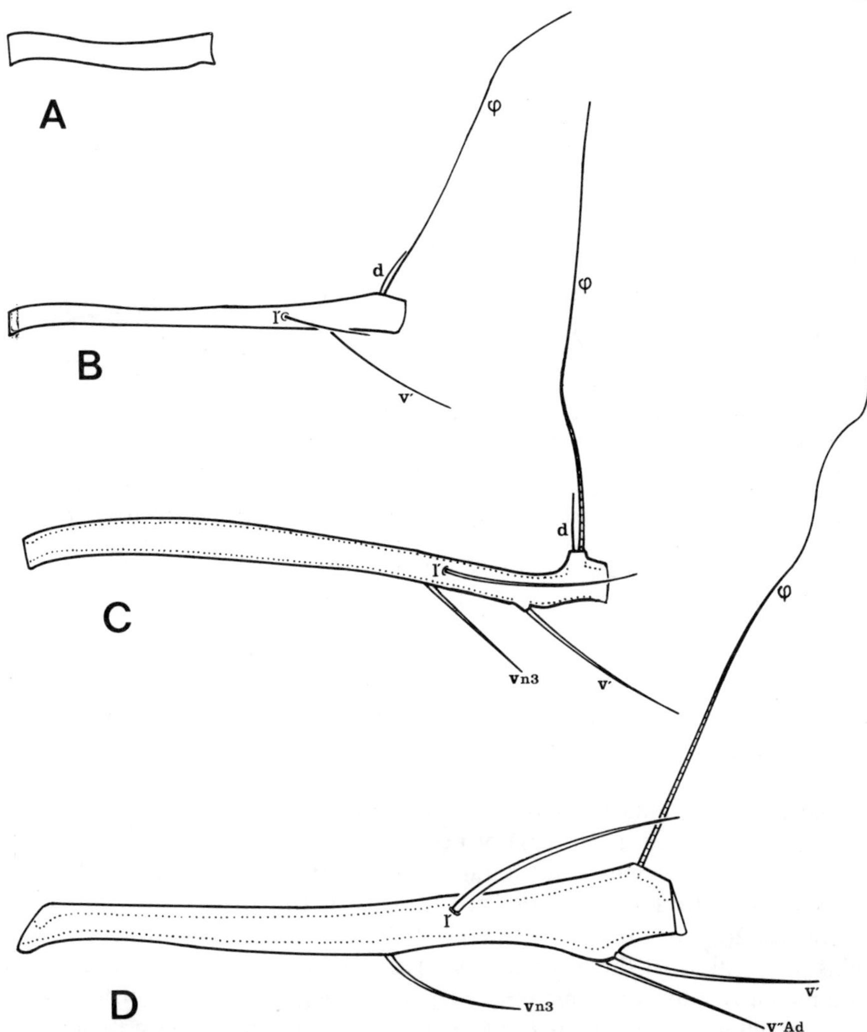

Fig. 2.10. Damaeus clavipes (Hermann), lateral (antiaxial = posterior) view of left tibia IV; A: protonymph; B: deutonymph; C: tritonymph; D: adult. An ontogenetic notation (the base level) is added to the homology notation in the case of the tritonymphal and adult phaneres. The protonymphal tibia is glabrous as a result of the protonymphal denudation of leg IV (a protelattosic regression which is a continuation of the complete regression of leg IV in the larva). Four phaneres (among which the dorsal seta *d*, the protector seta of the solenidion φ) are present in the deutonymph. Setae are added in the tritonymphal and in the adult stase. Seta *d*, present in deutonymph and tritonymph, is absent in the adult; it has disappeared as a result of the descendant regression of setae associated with a solenidion. The evolution of seta *d* is disharmonic (in the postembryonic ontogeny of *Damaeus clavipes* it is successively absent, present and again absent), because its disappearance is caused by two different evolutionary forces. A–D: × 232.

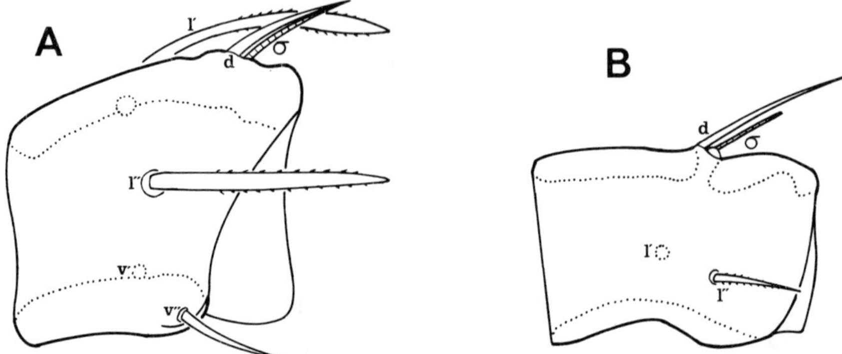

Fig. 2.11. Lateral (antiaxial) view of the genu of right leg II in the adult of two species of Nothroidea; A: *Nothrus silvestris* Nicolet; B: *Trimalaconothrus grandis* Van der Hammen. The greatest number of setae (the solenidion σ is no seta) on genu II, known from Nothroidea, is five (present in *Nothrus silvestris*); the smallest number known from Nothroidea is two (in the *Trimalaconothrus* species represented here, the number is three). The setae disappear, in the course of evolution, in a fixed order. This order is generally in accordance with the following priority list, in which the setae are arranged according to decreasing strength (the last-mentioned seta of the list is the first to disappear): (*d, l″*), *l′, v′, v″*. A, B: × 687.

often, but not always, also the most precocious in ontogeny.

A third approach pertains to the study of vertitions: the relative strength is often greater when the number of vertitional absences at a certain level of development is smaller, or when the number of vertitional presences at a certain level is greater.

There are a few indications that multiplication is a sign of weakness: the first seta to be multiplied, in the case of neotrichy, is sometimes also the weakest. This pertains, for example, to supracoxal setae *eII* of *Oecosmaris* and *Balaustium* (cf. Van der Hammen, 1977a; 13, Fig. 2 J, L).

The last approach (of which the results have a restricted applicability) pertains to the study of the metameric aspect: when two homonomous elements of a metamere have the same frequency, the stronger one is that of which the corresponding homonomous element of another metamere has a higher frequency. This approach can, for instance, be used in the case of appendicular organs. It should be applied with much care; it could, for instance, hold good in the case of legs I and II, but not in the case of legs III and IV.

Evidently, the homonomous elements of a group are organized or ranked in a hierarchy, which condition can be symbolized by a list of priority; this list can function, in the course of evolution, as an evolutionary program. As mentioned above, an organ of a group has a greater strength when it is more common and more precocious, when it presents less vertitional absences or more vertitional presences, when it is not multiplied, and when the corresponding homonomous element in other metameres is stronger. Generally, but not always, a study of each of these aspects leads to the same results.

In Fig. 2.11 the genu of right leg II of the adult of two species of Nothroidea is represented. The highest number of setae known for this segment in this group is five, the lowest number is two. The priority list, on which the setae are arranged according to decreasing strength (the last-mentioned seta of the list is the first to disappear) is (d, l''), l', v', v'' (the first two setae are of equal strength). In *Nothrus silvestris* Nicolet (Fig. 2.11 A) all setae are present. In *Trimalaconothrus grandis* Van der Hammen (Fig. 2.11 B) v'' and v' have disappeared. Other species are known in which v'' only, or v'', v' and l' are absent.

A comparative study of priority can be thwarted, when an element in a taxon is affected by more than one evolutionary change (cf. the disharmonic evolution mentioned at the end of section 5).

Another difficulty is presented by the study of accessory setae. In section 4 it was already mentioned that these are all of postlarval origin. Regarded as a group, they are weaker than the fundamental setae (all of larval origin). Within the group, however, the priorities seem to be conformable to particular rules.

In several taxa of primitive Oribatid mites, the priority lists for solenidia (sensory phaneres of which the sense is probably chemical) are not always in accordance with each other. This could be connected with the primitive condition (primordiotaxy) of the solenidiotaxy in question (cf. section 7).

The occurrence, in a taxonomic group, of more than one priority list, could be an indication of the artificial character of the group. Priority lists could constitute very important characters in a phylogenetic classification.

7. Evolution of setal patterns, with particular reference to dorsal, pleural and paraproctal opisthosomatic chaetotaxy

The subject dealt with in the present section has been studied, among others, by Grandjean (1934, 1934a, 1938, 1939b, 1942d, 1943, 1943c, 1947, 1949b, 1950a, 1952a, 1954, 1955, 1961a, 1963, 1965a), Coineau (1973, 1974), Van der Hammen (1975, 1979) and Travé (1979, 1979a). The survey given here is a further development of the views published by me before, in the two papers mentioned above.

A study of opisthosomatic chaetotaxy should preferably start from the original segmental arrangement. Opisthosomatic segmentation has been studied in primitive Actinotrichida (cf. Van der Hammen, 1970); the maximum number of opisthosomatic segments is ten (VII–XVI). The posterior three segments (XIV–XVI) are suppressed in the course of embryonic development (cf. section 5), so that in the larva the terminal segment is constituted by segment XIII (theoretically, segment XIII can also be suppressed). Dependent on the group, one or more of the suppressed segments can reappear in the course of postembryonic development (cf. Fig. 2.8).

The original segmental arrangement of dorsal, pleural and paraproctal opisthosomatic setae is still recognizable in many Actinotrichida. Usually, the paraproctal setae (i.e. the setae of segments XIII–XVI) and the notogastral setae (i.e. the dorsal, pleural and posterior setae of segments VII–XII in the larva, and of segments VII–XIII in nymphs and adult) are dealt with as two separate groups of setae (the setae of segment XIII are included in both groups). In the larva, the so-called holotrichous number of notogastral setae is thirteen pairs; in other stases the holotrichous number is sixteen pairs.

In some Palaeosomata (the most primitive group of Oribatid mites) higher numbers are, however, found. As long as these setae occur in fixed numbers, and are arranged in files, it is supposed that they represent also primitive setae; the condition is referred to as hypertrichy. Holotrichy as well as hypertrichy are cases of prototrichy (primitive chaetotaxies characterized by the exclusive occurrence, in a certain area, of idionymous[7] ancestral setae). Hypertrichy is also found in Endeostigmata (a primitive group of Actinedida). There is a gradual transition from hypertrichy to a more chaotic arrangement which will be dealt with below. Hypertrichy is characterized by a particular behaviour during postembryonic ontogeny: some of the hypertrichous setae (the so-called inguinal setae), present in the larva, disappear after the larval stase (generally, the total number of setae, in any given chaetotaxy, increases or remains unchanged in the course of postembryonic development).

In the course of evolution, the holotrichous number of setae can decrease by regression, resulting in deficiencies (meiotrichy). According to the number of pairs of suppressed setae, the deficient chaetotaxies are referred to as uni-, bi-, quadri- and multideficiency. The weakest of all holotrichous notogastral setae is constituted by seta f_1; it disappears by descendant regression (f_1 is sometimes still present in the larva). The evolution of deficiencies starts, as far as known, in a vertitional way.

In the course of evolution the number of setae can also increase. This increase must be attributed to a multiplication of the setae pre-existent in the area in question. This condition is called neotrichy. It is known, among others, from notogaster, epimera, genital and aggenital region, anal and adanal region, gnathosoma, and legs (neotrichy of the legs is known from various groups of Actinedida). It is interesting that in Trombidei with strongly developed neotrichy, the chelicerae have remained glabrous; apparently, all cheliceral setae had disappeared by regression before neotrichy arose, and consequently could no more be multiplied.

Travé (1979, 1979a) published summaries of all data with reference to neotrichy in Oribatid mites. Neotrichy appeared to occur more often in primitive than in higher Oribatid mites. An arrangement of the neotrichous parts of the body according to decreasing frequency (based on genera) resulted in the fol-

---

[7] Idionymy is defined (cf. section 2) as the quality of an organ of being capable of receiving a designation which is not collective, and enables a recognition among other homonomous organs, either in the course of ontogenetic development, or in a comparative study of a natural group. The antonym is adelonymy.

lowing list: aggenital region, epimera, notogaster, genital region, adanal region, anal region, and infracapitulum. In primitive Oribatid mites, neotrichy is found more often in the notogastral, the epimeral, and the genital region; in higher Oribatid mites, it is found more often in the aggenital and adanal region.

Grandjean (1947: 112) discovered that in Smarisidae the supracoxal seta *eII* (near the base of leg II) is duplicated in the course of ontogeny, whilst *eI* and *e* (the supracoxal setae of leg I and palp) remain simple. Of these three setae, *eII* is the weakest (the priority list is: *e, eI, eII*). From this it could be concluded that weak setae are multiplied before stronger setae. Apparently, strong setae have not only a greater resistance to regression, but also to multiplication. This could explain why neotrichy is more often found in primitive Oribatid mites, than in higher Oribatid mites (several of the weaker setae have disappeared in higher Oribatid mites). Consequently, parts of the body, most frequently affected by neotrichy, could be those which had preserved the weaker setae.

The evolution of neotrichy has not only phylogenetic, but also ontogenetic aspects. In the course of postembryonic development, neotrichy can be stationary (not increase after the stase of its first appearance) or growing (further increase after the stase of its first appearance).

It is possible to distinguish various types and degrees of neotrichy. A faintly developed neotrichy is called oligotrichy. A neotrichy with a distinctly recognizable, idionymous arrangement is called cosmiotrichy (when the setae are arranged in a file, it is called linear cosmiotrichy). When a single seta is multiplied in such a way that a tuft of setae takes the place of a single seta, neotrichy is called dragmatotrichy. A strongly developed neotrichy, characterized by a chaotic mass of setae, is called plethotrichy.

From investigations by Grandjean (1943c) and Coineau (1973, 1974) it has become evident that neotrichy starts locally, and can remain restricted to a small area. Coineau (1974) discovered that, in Caeculidae, the first beginnings of neotrichy are vertitional.

I have already tried before to prepare a model of the evolution of chaetotaxy, especially with reference to the dorsal, pleural and paraproctal region of the opisthosoma (Van der Hammen, 1975, 1979). The first serious problem which arises is constituted by primitive chaetotaxy. I have mentioned above that primitive holotrichy is preceded by the still more primitive hypertrichy. When setae in primitive taxa are more numerous, it is very difficult to distinguish hypertrichy from neotrichy (especially linear cosmiotrichy). Hypertrichy is known from primitive Palaeosomata (Oribatida) and from primitive Endeostigmata (Actinedida). As mentioned above, the behaviour of some hypertrichous setae (the inguinal setae), in the course of postembryonic ontogeny, is very interesting: they disappear after the larval stase (just as seta f_1 in several species). Apparently, a number of the most primitive setae have disappeared (or are in course of disappearance) by descendant regression.

Studies in opisthosomatic chaetotaxy in Opilioacarida, and leg chaetotaxy

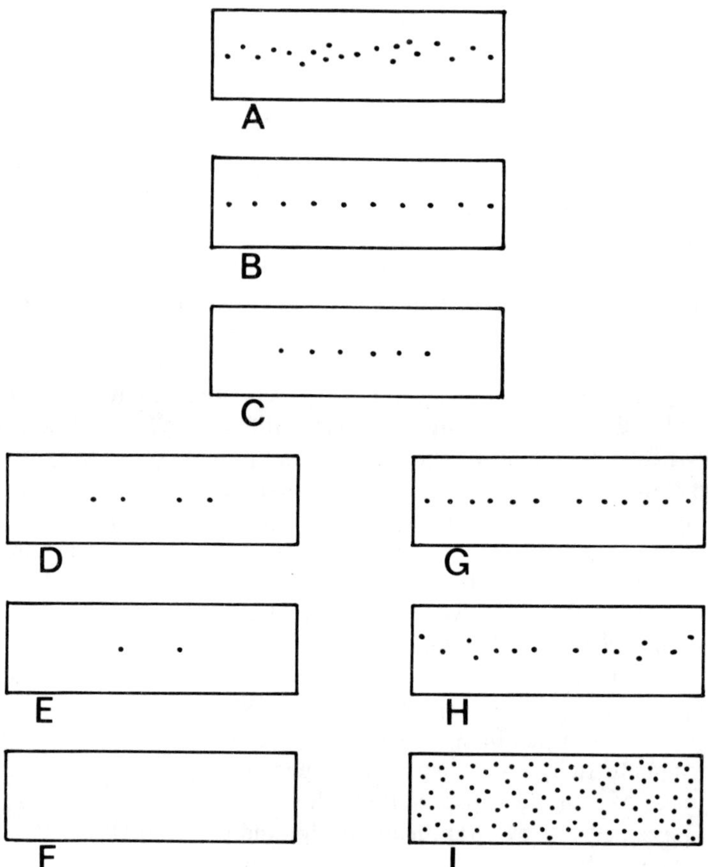

Fig. 2.12. Schematic representation (a rectangle representing a segment) of the various stages in the evolution of chaetotaxy in Actinotrichida; A: primordiotrichy; B: hypertrichy; C: holotrichy; D, E: meiotrichy; F: atrichosy; G–I: neotrichy; G: cosmiotrichy; H: oligotrichy; I: plethotrichy. Primordiotrichy and hypertrichy are not (or not yet, or not always) distinguishable from neotrichy.

in Holothyrida (both taxa constitute primitive groups of Anactinotrichid mites) led me to introduce a new hypothesis with reference to the evolution of chaetotaxy. I supposed that primitive prototrichy (hypertrichy and holotrichy) is preceded, in evolution, by a still more primitive, more numerous and more chaotic chaetotaxy. This hypothetic condition was named primordiotrichy. I pointed out the fact that, according to our present knowledge, primordiotrichy is not distinguishable from neotrichy. Prototrichy could have arisen from primordiotrichy by a special type of numerical regression (an evolutionary process called trichodelosis). In this case regression could refer, not to suppression of individual setae, but to the development of individually coded setae from setae which, in the genome, were collectively or partly collectively coded.

Primitive prototrichy can be changed under the influence of two evolutionary processes: (1) regression and the development of deficiencies (trichomeiosis, leading to meiotrichy) and (2) multiplication (neotrichosis, leading to neotrichy). Neotrichy can, of course, also arise from meiotrichy. Trichomeiosis as well as neotrichosis are evolutionary processes which start in a vertitional way and are, evidently, controlled by changes in systems of interactions, relative strength, and distributional information.

A schematic survey of the various types of chaetotaxy is given in Fig. 2.12. The types could be classified in two groups: (1) atactotrichy, a chaotic chaetotaxy, comprising primordiotrichy and adelonymous neotrichy; and (2) idiotrichy, comprising primitive prototrichy, meiotrichy, and idionymous neotrichy.

8. Ontophylogenetic aspects of genital chaetotaxy in Oribatida

The genital setae of Oribatid mites are inserted on the progenital covers (or lips or valves). As mentioned in section 1, the covers are supposed to belong to two segments (VIII−IX) and to represent the fused exites of opisthosomatic appendages (Van der Hammen, 1982: 31−34, Fig. 26). The endites of these opisthosomatic segments surround the eugenital opening, whilst the opisthosomatic appendages of segments X−XII are represented by the genital papillae (inside the progenital chamber, covered by the progenital lips or valves); the number of pairs of genital papillae in proto-, deuto- and tritonymph, and adult is 1-2-3-3 (in normal cases).

The development of the progenital region (progenital lips or valves, genital papillae and genital setae) starts in the protonymph; genital setae can be added by each of the subsequent stases (deutonymph, tritonymph, adult) (cf. Fig. 2.13). The development of the genital chaetotaxy can be expressed in a formula consisting of four numbers (the numbers of genital setae of respectively protonymph, deutonymph, tritonymph and adult).

A comparative study enables us to distinguish a primitive formula (1-5-8-10), formulae which are the result of numerical regression, and formulae representing neotrichy. There are several formulae which can be attributed to regressive evolution. These formulae range from the primitive formula (1-5-8-10) to (1-1-1-1). The prototrichous formulae all start with the number 1 (for the protonymph).

Grandjean (1961b) developed a hypothetic model for the regressive evolution of the genital setae. In his observations on the ontogenetic aspect, he distinguished two cases, assuming that the setae are independent of each other: eustasy (in which each seta appears at a fixed stase, unless it is absent) and amphistasy (in which each seta can appear at any stase). In the hypothetical model, the genital setae are supposed to be generally eustasic (otherwise the

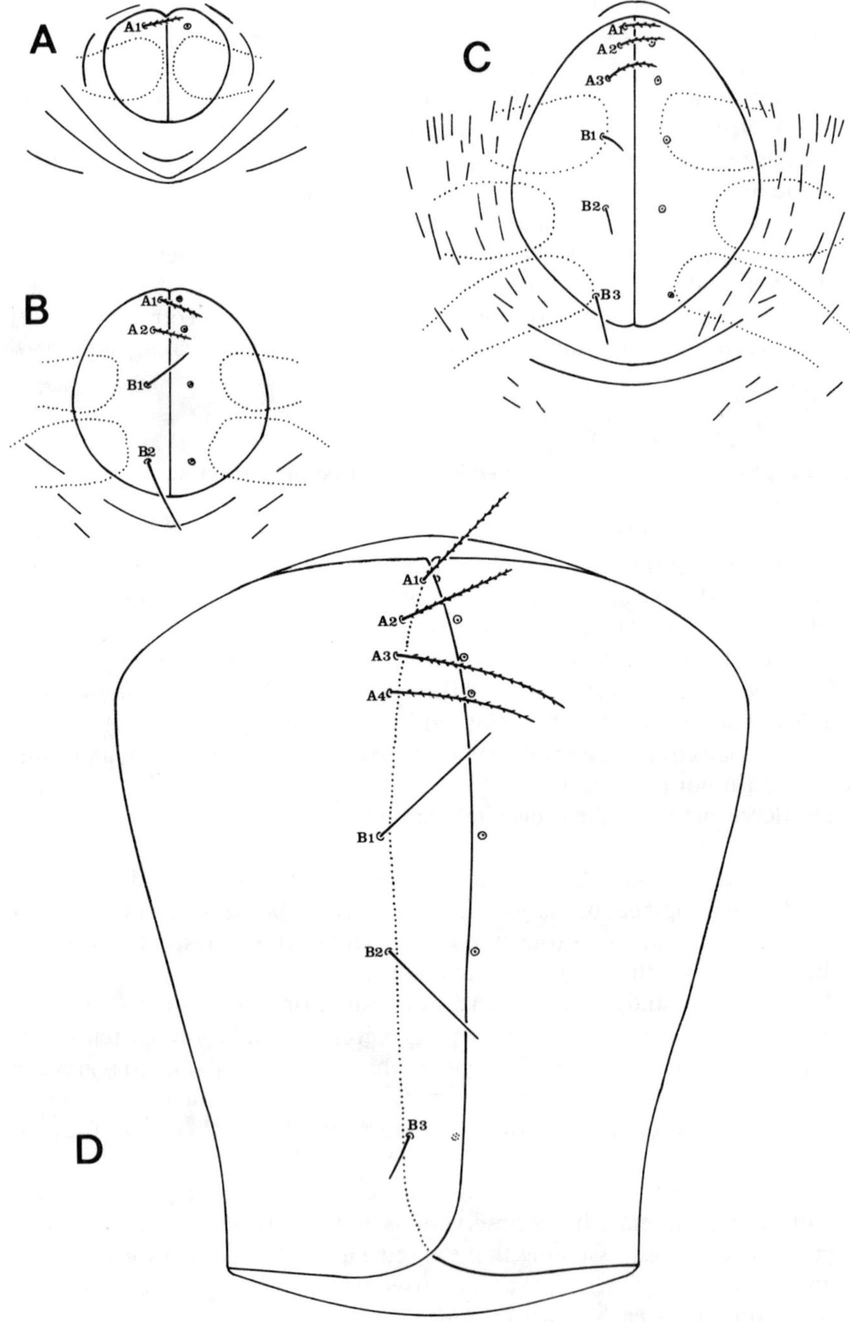

number of known formulae would be considerably greater) and to be suppressed in a given regular sequence. Starting from these assumptions as well as from the formula (1-1-1-1) and the fact that all prototrichous formulae start with the number 1, Grandjean tried to reconstruct a priority list. The list prepared by him did not permit of a homologization of setae, but still led to important conclusions. Apparently, there are stages in the regression, which are difficult to pass, and which are reflected by the most common formulae (the number of formulae found in nature is, even in the case of eustasy, much smaller than the number of formulae which are theoretically possible). One of the most common formulae is, for instance, the formula (1-3-5-6). It is assumed that the common formulae (which represent the stages of detainment) are caused by the unequal resistance of the setae to regression.[8] The formula (1-3-5-6), for instance, would be determined by the greater strength of the second deutonymphal seta (according to the argumentation given below, and according to the notation of Fig. 2.13, this would be seta *B1*).

The hypothetical list of priority cannot account for a small number of deviating formulae (found in some primitive Oribatid mites). These formulae could perhaps be interpreted as the result of amphistasy.

In *Archegozetes magna* (Sellnick), of which the genital region and its development are represented in Fig. 2.13, two types of genital setae are found: ciliate setae and smooth setae. This condition enables a further development of Grandjean's model, introducing the segmental arrangement of the setae. All ciliate genital setae (*A1 – A4*) of *Archegozetes magna* apparently belong to segment VIII, all smooth setae (*B1, B2, B3*) to segment IX. *A1* is the single protonymphal seta; *A2, A3* and *A4* are added by deutonymph, tritonymph and adult respectively. Apparently, the genital chaetotaxy of segment VIII has a normal development, and has not been subject to regression. The genital chaetotaxy of segment IX starts, in the deutonymph, with two setae (*B1* and *B2*), after which one seta (*B3*) is added by the tritonymph, and no seta by the adult. Apparently, at least one seta (*B4*) is suppressed. Evidently, the genital setae of segment VIII are stronger than those of segment IX, whilst the nymphal setae are stronger than the seta with adult base level. The results are interesting enough to try to find other cases with more than one type of genital setae, in order to compare the development of these with that of *Archegozetes magna*.

The numerical evolution of genital chaetotaxy can also be progressive, resulting in genital neotrichy. This is (in the case of Oribatid mites), for in-

---

[8] Regression in number could be associated with an increase in strength of the remaining elements. This could perhaps partly explain a stage of detainment.

←

Fig. 2.13. Postembryonic development of the genital region in *Archegozetes magna* (Sellnick); A: protonymph; B: deutonymph; C: tritonymph; D: adult; A–D: × 475.

stance, known from several Nothroidea and from the genera *Hydrozetes, Limnozetes* and *Niphocepheus*.

Two types of genital neotrichy can be distinguished: stationary neotrichy and growing neotrichy. Stationary neotrichy is, for instance, known from species of the genera *Hydrozetes* and *Limnozetes*, in which species one extra seta only is added (at the level of the protonymph). The genital formula in question is (2-4-6-7), which can be derived from (1-3-5-6), found in other species of these genera.

In most known cases, neotrichy is growing: the number of genital setae increases in subsequent stases, in a way different from the prototrichous formulae. The genital formula of *Trimalaconothrus novus* (Sellnick), for instance, is 2-(3-4)-(6-7)-(8-12).

The most interesting aspects of the study of genital chaetotaxy are not only constituted by the models for priority and regression, but also, and especially, by the discovery of the probable existence of stages of detainment, which are difficult to pass. Evidently, unknown resistances (related to segmentation?) must be overcome, before regression can again proceed.

9. Ontophylogenetic aspects of solenidiotaxy in Oribatida

Solenidia are hollow sensory phaneres with thin walls (with pores), which could be chemosensory. They resemble setae, but have a very short, largely open root. In Actinotrichida, they can be found on the palpal tarsus, and on femur, genu, tibia and tarsus of the legs. In Oribatida, solenidia do not occur on the femur of the legs. The present section deals with ontophylogenetic aspects of solenidiotaxy in the legs of Oribatids (Fig. 2.14). These aspects have been studied by Grandjean (1935, 1946, 1964).

The occurrence and distribution of solenidia can be expressed in formulae, in which the numbers for the three leg segments are mentioned for the four legs. In *Nothrus palustris* C.L. Koch (Fig. 2.14) the solenidial formula of the adult is: (1-2-3) (1-1-1) (1-1-0) (1-1-0).

Solenidial formulae can be prepared for all stases. Many solenidia are of larval origin, but other solenidia can be added in subsequent stases. In the case of *Nothrus palustris*, $\sigma_1 I$, $\varphi_1 I$, $\omega_1 I$, $\sigma_1 II$, $\varphi_1 II$, $\omega_1 II$, $\sigma_1 III$ and $\varphi_1 III$ are of larval origin; $\omega_2 I$ is of protonymphal origin; $\varphi_2 I$, $\sigma_1 IV$ and $\varphi_1 IV$ are of deutonymphal origin; $\omega_3 I$ is of tritonymphal origin.

In the course of evolution, the number of solenidia has, in many cases, been subject to numerical regression. When solenidia are arranged in a list, according to the order of ontogenetic appearance, i.e. in the order of decreasing strength, they constitute an ontogenetic priority list. Such a list must fit all species of a natural group.

The aleatory character of a solenidion (i.e. its absence in some specimens, often at one side only) is a sign of weakness; the absence is vertitional. The oc-

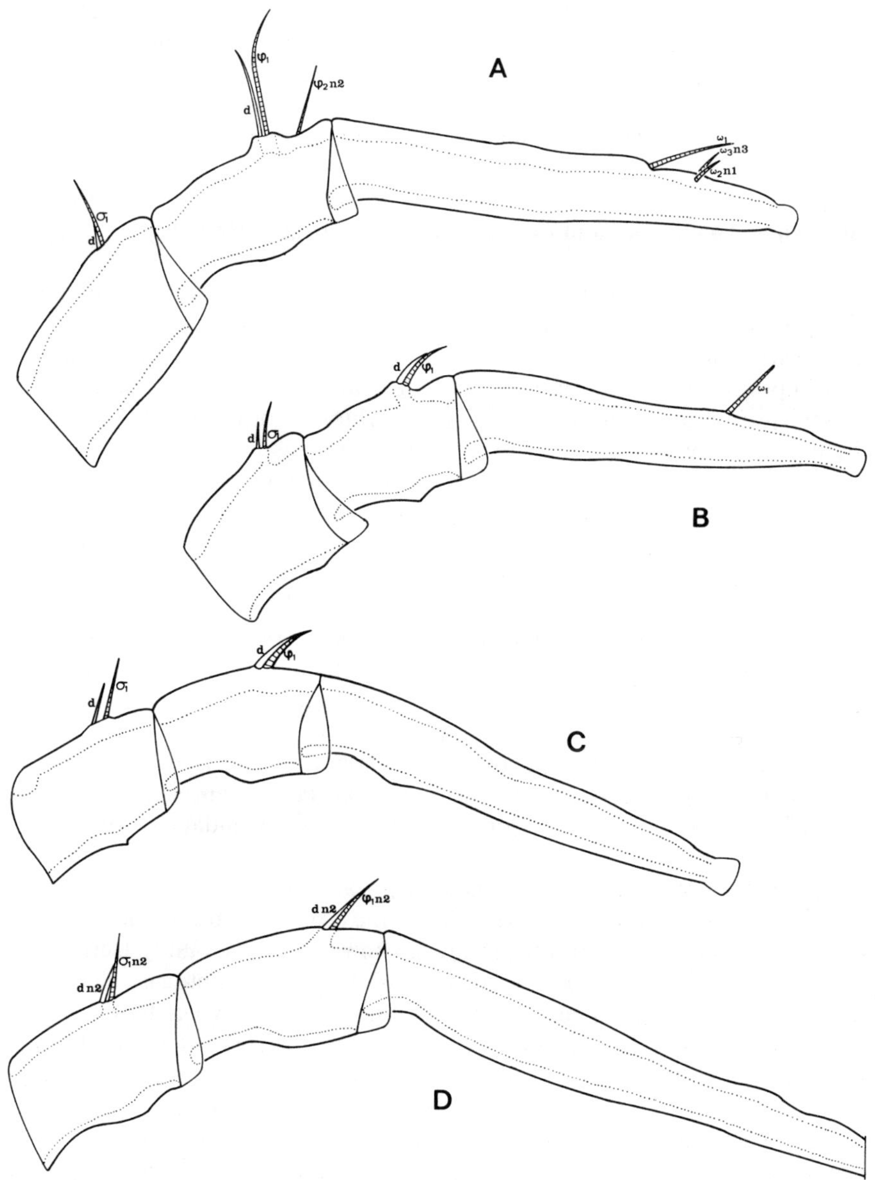

Fig. 2.14. Solenidiotaxy of legs I–IV of *Nothrus palustris* C.L. Koch (all setae omitted, with the exception of the protector setae *d*); genu, tibia and tarsus of the adult right legs are represented, and the base levels of solenidia and protector setae are added to the homology notation (no base level is mentioned in case the phanere is of larval origin); A–D (legs I–IV): × 246.

54

currence of amphistasy (cf. section 8) constitutes also a sign of weakness. Several stages of detainment (represented by the common formulae) can again be recognized; they are caused by the greater resistance to regression of some solenidia. Here again, important resistances must apparently be overcome, before regression can proceed, and pass a stage of detainment.

10. Numerical changes and evolution in Nothroidea (Oribatida)

Many of the numerical changes, discussed in the previous sections, manifest themselves at the same time, for instance in a single natural group and in a single species. For a better understanding of the systems underlying these changes, it is important to prepare a survey of this simultaneous occurrence. In the present section the numerical changes occurring in Nothroidea, are therefore discussed; many relevant data with reference to this group are known. The superfamily Nothroidea constitutes one of the higher groups of primitive Oribatid mites. Seven families are now distinguished, viz., Nothridae, Crotoniidae, Camisiidae, Trhypochthoniidae, Malaconothridae, Nanhermanniidae and Hermanniidae. In the present section the numerical changes will be discussed in two ways: arranged according to body regions and appendages; and arranged according to ontophylogenetic behaviour. In the morphological approach, the numerical changes are discussed under nine headings (cf. Figs. 2.1, 2.2): prodorsum, notogaster, paraproctal region, genital region, aggenital region, epimera, gnathosoma, palp, and legs.

Among the prodorsal elements, the trichobothrium has been subject to regression in most of the families; the regression is ascendant. One or both exobothridial setae can have disappeared, either by descendant or by vertical regression.

Among the notogastral setae, f_1 has disappeared in most families by descendant regression; f_2 and h_3 can be lacking in the larva, and are subject to ascendant regression. Notogastral neotrichy is known from the family Hermanniidae (Travé, 1979a: 594). The occurrence of 17 pairs of notogastral setae in *Austronothrus* (Crotoniidae) could constitute a case of hypertrichy.

The adanal and anal segments are, in all members of the superfamily, suppressed in the course of embryonic development, and appear respectively in the protonymph (adanal segment) and the deutonymph (anal segment); in ontophylogenetic diagrams for these regressions, the line of chronological separation is vertical. The appearance of the paraproctal setae is often retarded by one stase (paraproctal atrichosy, either at two or at three levels; cf. Fig. 2.8); in this case, a similar agent is successively affecting a series of heteronomous regions close to the anal opening, and not a single idionymous element. One of the adanal setae, and one or two of the anal setae can disappear by a regression which is apparently ascendant.

Several genital setae can be suppressed; the setae are supposed to be eustasic, and the regression is apparently vertical. Genital neotrichy is found in several groups of Nothroidea; it is either stationary (as in *Hermannia convexa* (C.L. Koch), in which one seta is added in the protonymph) or growing (as in many Camisiidae, Trhypochthoniidae and Malaconothridae).

The aggenital setae can be subject to regression as well as to multiplication. They could be eustasic (in which case regression would be vertical); otherwise they could disappear by ascendant regression. Aggenital neotrichy is known from the genus *Hermannia*.

The epimeral setae can also be subject to regression as well as to multiplication. Regression can be vertitional and ascendant, although many epimeral setae could be eustasic. Epimeral neotrichy is known from Nothridae, Crotoniidae, Nanhermanniidae and Hermanniidae. It is either stationary (*Holonothrus*) or growing. It pertains mostly to epimera 4 and 3, although epimera 1 and 2 can also be affected (*Nothrus*; cf. Sellnick & Forsslund, 1955: 496).

The median infracapitular setae of the gnathosoma are subject to regression; in the case of m_2, the regression is known to be ascendant. The number of adoral setae can also be subject to regression (ontophylogenetic type unknown).

Several setae of the palp are subject to regression, notably setae of femur, genu, tibia and tarsus. In several cases the regression is known to be ascendant.

Many setae and solenidia of the legs, as well as the lateral ungues of the claws, can be subject to regression. In several cases the regressions are known to be vertitional and ascendant. Several setae and solenidia are known to be eustasic; in these cases regression is vertical.

As mentioned above, our second approach to the numerical changes in Nothroidea is ontophylogenetic. Regressions and multiplications are discussed here under five headings: ascendant regression, descendant regression, vertical regression, stationary neotrichy and growing neotrichy.

Ascendant regression is common in Nothroidea; it starts with retardation and ends with complete suppression. It is, for instance, known from: the trichobothrium; the notogastral setae f_2 and h_3; the median infracapitular seta m_2; several setae of the palp; several setae and solenidia of the legs, and the lateral ungues of the claws.

Descendant regression is rare in Nothroidea. It is known from notogastral seta f_1. In some Trhypochthoniidae, the single exobothridial seta, present in the larva, is suppressed in the protonymph and subsequent stases, probably as a result of descendant regression.

Vertical regression is known from the paraproctal segments and from other eustasic elements (although several eustasic phaneres could have been amphistasic in an earlier period of evolution).

Stationary neotrichy is known from the genital setae of *Hermannia convexa* (C.L. Koch) and the epimeral setae of *Holonothrus foliatus* Wallwork.

Growing neotrichy is known from the genital region (Camisiidae, Trhypochthoniidae, Malaconothridae), the aggenital region (Hermanniidae) and the

56

epimera (Nothridae, Crotoniidae, Nanhermannidae and Hermannidae).

Although the data mentioned in the present section are interesting, and demonstrate the simultaneous occurrence of various types of numerical changes, it is perhaps still too early for the preparation of a general model. An information processing approach, however, could perhaps disclose interesting aspects of the underlying systems of interactions (epigenetic and evolutionary codes).

Conclusions

The results, obtained in the preceding sections, enable us to bring together some of the data required for the preparation of a model of the numerical evolution of setae, solenidia, segments, etc. An evolution of this type starts vertitionally, by manifestation of a change in a system of genotypic interactions (at first in a small number of specimens of a population; the change is not hereditary, because the offspring inherits only the evolutionary potentiality for the change and a certain probability of manifestation); the statistical probability of manifestation is at first small, and can result in seemingly irregular, although balanced patterns of the change (distributional information on the whole is evidently present in the parts). It has appeared that the manifestation of a vertitional evolution can be the result of a co-ordination of changes in the environment and in the systems of genotypic interactions. Identical systems of interactions operate in large taxonomic groups. Many evolutionary changes manifest themselves in the course of postembryonic development, and changes in the ontogenetic aspect of evolution are apparently manifestations of changes in systems of genotypic interactions.

In simple cases, there are three different ways in which numerical evolutionary changes manifest themselves in the course of postembryonic development: the change can first manifest itself in one of the early instars and subsequently affect the later instars; the change can first manifest itself in one of the later instars and subsequently affect the earlier instars; and the change can always manifest itself at the same level of postembryonic development (it is possible that this so-called eustasy of an element is preceded, in the course of evolution, by amphistasy, i.e. variability in the level of manifestation).

It has appeared that different changes in systems of interactions can affect one and the same element (changes in the character states of homologous elements, consequently, can be heterologous). It has also appeared that homonomous elements generally present differences in the resistance to suppression and multiplication, which condition can be symbolized by a list of priority; among these differences, there are more important ones that can result in certain stages of detainment (which are difficult to pass).

The model of chaetotactic evolution, prepared in section 7 (Fig. 2.12), could perhaps be developed into a general model of numerical evolution, by distinguishing three stages: (a) a primordiotactic stage, in which groups of homonomous elements were collectively coded; (b) a prototactic stage, in which homonomous elements (in fundamental numbers) are individually coded[9]; and (c) an alassotactic stage (or period) in which the prototactic numbers of elements are either reduced (meiotaxy) or multiplied (neotaxy). Coding could, ancestrally, have been segmentally organized. The above-mentioned model can also be applied to the numerical evolution of the chelicerate life-cycle (cf. Van der Hammen, 1978, and essay III).

The discovery of the existence of evolutionary potentialities in the genotype and of a coded order in the manifestation of evolution, could constitute an important contribution to phylogenetic systematics. Evidently, the evolutionary program itself constitutes a much better taxonomic character than the manifestation of one of the coded changes. The existence of evolutionary potentialities in the genotype could explain many parallelisms in evolution (cf. Van der Hammen, 1986a, and essay IV).

As mentioned at the end of section 10, an information processing approach to numerical evolution could perhaps disclose interesting aspects of the underlying systems of interactions (epigenetic and evolutionary codes). These systems are certainly characterized by a hierarchical structure, corresponding with the hierarchy of the natural system; one of the deepest levels of this structure probably pertains to the origin of segmentation. In the course of evolution, fundamental differences between taxa arose by the origination of different evolutionary programs (and not by the manifestation of the changes in the phenotype). The study of the numerical evolution of setae and setiform organs suggests that the structural genes for these elements are similar and completely present in all Actinotrichida and that differences between the groups arose as a result of changes in hierarchical systems of interactions.

It may finally be repeated here (cf. essay I, subsection 6.3) that decisions on chaetotactic patterning (and its postembryonic development) are apparently made very early in embryonic development. It is, apparently, in this way that a balanced distribution of vertitional changes comes into being, even in the case of patterns which manifest themselves only after repeated postembryonic moulting (as in the case of variation in the number of adult ungues).

---

[9] In Chelicerata, four different prototactic numbers of segments probably arose from the 'prochelicerate' primordiotactic (and perhaps also anamorphic) segmentation: nineteen (Cryptognomae, Scorpionida), eighteen (Arachnida s. str.), seventeen (Epimerata, Apatellata) and sixteen (Opilionida, Xiphosura). As mentioned in the introduction and in section 5, the fundamental number of segments in Actinotrichida is reduced by suppression during embryonic development, and partial reappearance during postembryonic development (hysteromorphosis). The numbers of segments apparently constitute important characteristics of the groups (although several groups independently developed the same prototactic numbers).

58

List of the notations of Figs. 2.1–2.14.

a	anterior infracapitular seta.
a', a'', (a)	antelateral tarsal setae.
$A1$, $A2$, $A3$, $A4$	genital setae attributed to segment VIII.
ad, Ad	adult, adult base level.
AD	adanal segment (segment XIV).
ad_1, ad_2, ad_3	adanal setae.
ag_1, ag_2	aggenital setae.
AN	anal segment (segment XV).
an_1, an_2	anal setae.
AP	apotele.
$a.p$	area porosa.
$B1$, $B2$, $B3$	genital setae attributed to segment IX.
bl'	anterior basilateral seta of femur.
bo	bothridium.
bv''	posterior basiventral seta of femur.
c_1, c_2, c_3, c_p	notogastral setae attributed to segments VII and VIII.
d	dorsal seta of leg-segment.
d_1, d_2	notogastral setae attributed to segment IX.
e_1, e_2	notogastral setae attributed to segment X.
ex	exobothridial seta.
F	femur.
f_1, f_2	notogastral setae attributed to segment XI.
ft', ft'', (ft)	fastigial setae of tarsus.
GE	genu.
GEN	genital valves.
gla	latero-opisthosomatic gland.
h	posterior infracapitular seta.
h_1, h_2, h_3	notogastral setae attributed to segment XII.
iad	lyrifissure or cupule of the adanal segment (segment XIV).
ian	lyrifissure or cupule of the anal segment (segment XV).
ih	lyrifissure or cupule attributed to segment XII.
im	median lyrifissure of notogaster (attributed to segment X).
in	interlamellar seta.
ip	posterior lyrifissure of notogaster (attributed to segment XI).
ips	lyrifissure or cupule of the pseudanal segment (segment XIII).
it', it''	iteral setae of tarsus.
l', l'', (l)	lateral setae of legs.
le	lamellar seta.
lv	larva.
m_1, m_2	median infracapitular setae.
$n1$	protonymph, protonymphal base level.

n2	deutonymph, deutonymphal base level.
n3	tritonymph, tritonymphal base level.
oc	central unguis.
ol', ol"	lateral ungues.
P	ancestral character state.
(p)	proral setae of tarsus.
pl', pl"	primilateral setae of tarsus.
plv	prelarva.
PS	pseudanal segment (segment XIII).
ps_1, ps_2, ps_3	pseudanal setae.
pv', pv"	primiventral setae of tarsus.
ro	rostral seta.
s	subungual seta of tarsus.
S	derived character of state.
se	sensillus (bothridial seta).
t	ontogenetic time.
T	phylogenetic time.
TA	tarsus.
tc', tc", (tc)	tectal setae of tarsus.
TI	tibia.
TR	trochanter.
(u)	ungual setae of tarsus.
v', v" (v)	ventral setae of legs.
ϵ	famulus of tarsus I.
ζ	eupathidium.
σ, σ_1	solenidia of genu.
$\varphi, \varphi_1, \varphi_2$	solenidia of tibia.
$\omega_1, \omega_2, \omega_3$	solenidia of tarsus.
I, II, III, IV	legs I–IV.
XIII	pseudanal segment.
XIV	adanal segment.
XV	anal segment.
1a, 1b, 1c	epimeral setae of segment III.
2a	epimeral seta of segment IV.
3a, 3b, 3c	epimeral setae of segment V.
4a, 4b, 4c, 4d	epimeral setae of segment VI.
'	(prime), anterior element of legs (paraxial in the case of legs I–II, antiaxial in the case of legs III–IV).
"	(double prime), posterior element of legs (antiaxial in the case of legs I–II, paraxial in the case of legs III–IV).
( )	a pair (of setae).

References

Beer, G. de, 1958. Embryos and ancestors. – Oxford (University Press): xii + 197 p., Figs. 1–19 (third edition).

Boelé, F.E. & L. van der Hammen, 1982. Variation in the number of ungues in Ameronothrus schneideri (Oudemans, 1903), an Oribatid mite. – Zool. Meded. Leiden 56: 169–191, Fig. 1, Tab. 1–5.

Coineau, Y., 1973. Ontophylogenetic considerations on the chaetotaxy of Caeculidae (Acari-Prostigmata). – Proc. 3rd Int. Congr. Acarol. Prague 1971: 269–273, Figs. 1–3.

Coineau, Y., 1974. Eléments pour une monographie morphologique, écologique et biologique des Caeculidae (Acariens). – Mém. Mus. Nat. Hist. Natur. (N.S.), sér. A, Zool. 81: i–vi, 1–299, Figs. 1–115, Pls. 1–24.

Dobzhansky, Th., F.J. Ayala, G.L. Stebbins & J.W. Valentine, 1977. Evolution. – San Francisco (W.H. Freeman and Company): xiv + 572.

Elder, D., 1979. An epigenetic code. – Differentiation 14: 119–122, Figs. 1–2, Tab. 1–3.

Elder, D., 1984. Theory of epigenetic coding. – Journ. Theor. Biol. 108: 327–332.

García-Bellido, A., 1983. Comparative anatomy of cuticular patterns in the genus Drosophila. – In: Development and Evolution (Cambridge): 227–255, Figs. 1–12.

Grandjean, F., 1934. La notation des poils gastronotiques et des poils dorsaux du propodosoma chez les Oribates (Acariens). – Bull. Soc. Zool. France 59: 12–44, Figs. 1–10. [= Œuvres, I(19)]. (The numbers are those of the Œuvres Acarologiques Complètes (Complete Acarological Works), vols. I–VII, edited by L. van der Hammen (The Hague, Lochem, 1972–1976)).

Grandjean, F., 1934a. Oribates de l'Afrique du Nord (2me série). – Bull. Soc. Hist. Natur. Afr. Nord 25: 235–252, Figs. 1–5. [= Œuvres I(23)].

Grandjean, F., 1935. Les poils et les organes sensitifs portés par les pattes et le palpe chez les Oribates. – Bull. Soc. Zool. France 60: 6–39, Figs. 1–8. [= Œuvres I(26)].

Grandjean, F., 1938. Observations sur les Acariens (4e série). – Bull. Mus. Nat. Hist. Natur. (2) 10: 64–71, Figs. 1–3. [= Œuvres II(46)].

Grandjean, F., 1938a. La suppression d'organes dans l'évolution d'une série homéotype. – C.R. Séanc. Acad. Sci. 206: 1853–1856 (sep. p. 1–5). [Œuvres II(50)].

Grandjean, F., 1939. La répartition asymétrique des organes aléatoires. – C.R. Séanc. Acad. Sci. 208: 861–865 (sep. p. 1–4). [= Œuvres II(56)].

Grandjean, F., 1939a. Au sujet de la probabilité d'existence des organes et des caractères. – C.R. Séanc. Acad. Sci. 208: 1456–1460 (sep. p. 1–4). [= Œuvres II(59)].

Grandjean, F., 1939b. Quelques genres d'Acariens appartenant au groupe des Endeostigmata. – Ann. Sci. Natur., Zool. (11) 2: 1–22, Figs. 1–25. [= Œuvres II(60)].

Grandjean, F., 1939c. L'utilisation des écarts dans les problèmes d'homologie que pose le développement discontinu des organes homéotypes. – C.R. Séanc. Acad. Sci. 209: 814–818 (sep. p. 1–4). [= Œuvres, II(63)].

Grandjean, F., 1939d. Les segments post-larvaires de l'hystérosoma chez les Oribates (Acariens). – Bull. Soc. Zool. France 64:. 273–284, Figs. 1–4. [= Œuvres II(64)].

Grandjean, F., 1939e. L'évolution des ongles chez les Oribates (Acariens). – Bull. Mus. Nat. Hist. Natur. (2) 11: 539–546. [= Œuvres II(65)].

Grandjean, F., 1941. La chaetotaxie comparée des pattes chez les Oribates (1re série). – Bull. Soc. Zool. France 66: 33–50, Figs. 1–4. [Œuvres II(70)].

Grandjean, F., 1941a. Sur la priorité dans les groupes d'organes homéotypes qui évoluent par tout ou rien. – C.R. Séanc. Acad. Sci. 213: 417–421. (sep. p. 1–5). [Œuvres II(72)].

Grandjean, F., 1942. Les méthodes pour établir des listes de priorité et la concordance de leurs résultats. – C.R. Séanc. Acad. Sci. 214: 729–733 (sep. p. 1–4). [Œuvres II(75)].

Grandjean, F., 1942a. La chaetotaxie comparée des pattes chez les Oribates (2e série). – Bull. Soc. Zool. France 67: 40–53, Figs. 1–2. [Œuvres II(78)].

Grandjean, F., 1942b. La signification évolutive des écarts individuels. – C.R. Séanc. Acad. Sci. 215: 216–220 (sep. p. 1–4). [Œuvres II(80)].

Grandjean, F., 1942c. Sur deux types généraux de régression numérique chez les Acariens. –

C.R. Séanc. Acad. Sci. 215: 236–239 (sep. p. 1–4). [Œuvres II(81)].

Grandjean, F., 1942d. Quelques genres d'acariens appartenant au groupe des Endeostigmata (2e série). – Première partie. – Ann. Sci. Natur. Zool. (11) 4: 1–135, Figs. 1–3. [Œuvres III(95)].

Grandjean, F., 1943. L'orthotaxie, la pléthotaxie et les écarts en biologie. – C.R. Séanc. Soc. Phys. Hist. Natur. Genève 60: 118–123. [Œuvres II(82)].

Grandjean, F., 1943a. Priorité absolue et statistique en biologie. – C.R. Séanc. Soc. Phys. Hist. Natur. Genève 60: 135–139. [Œuvres II(83)].

Grandjean, F., 1943b. La probabilité des organes en biologie. – Bull. Mus. Nat. Hist. Natur. (2) 15: 175–180. [Œuvres II(86)].

Grandjean, F., 1943c. Le développement postlarvaire d' 'Anystis' (Acarien). – Mém. Mus. Nat. Hist. Natur. (N.S.) 18: 33–77, Figs. 1–13. [Œuvres II(87)].

Grandjean, F., 1946. Les poils et les organes sensitifs portés par les pattes et le palpe chez les Oribates. Troisième partie. – Bull. Soc. Zool. France 71: 10–29. [Œuvres III(98)].

Grandjean, F., 1946a. La signification évolutive de quelques caractères des Acariens (1re série). – Bull. Biol. France Belg. 79: 297–325, Figs. 1–3. [Œuvres III(99)].

Grandjean, F., 1947. Etude sur les Smarisidae et quelques autres Erythroïdes (Acariens). – Arch. Zool. Exp. Gén. 85: 1–126, Figs. 1–11. [Œuvres III(104)].

Grandjean, F., 1947a. Sur la distinction de deux sortes de temps en biologie évolutive et sur l'attribution d'une phylogenèse particulière à chaque état statique de l'ontogenèse. – C.R. Séanc. Acad. Sci. 225: 612–615 (sep. p. 1–4). [Œuvres III(106)].

Grandjean, F., 1947b. L'harmonie et la dysharmonie chronologique dans l'évolution des stases. – C.R. Séanc. Acad. Sci. 225: 1047–1050 (sep. p. 1–4). [Œuvres III(107)].

Grandjean, F., 1948. Sur l'élevage de certains Oribates en vue d'obtenir des clones. – Bull. Mus. Nat. Hist. Natur. (2) 20: 450–457. [Œuvres III(114)].

Grandjean, F., 1948a. Sur les écarts dans un clone de Platynothrus peltifer (Acarien). – C.R. Séanc. Acad. Sci. 227: 658–661 (sep. p. 1–4). [Œuvres III(115)].

Grandjean, F., 1948b. Sur le phénomène des écarts. – C.R. Séanc. Acad. Sci. 227: 879–882 (sep. p. 1–4). [Œuvres III(116)].

Grandjean, F., 1949. Sur les rapports théoriques entre écarts et mutations. – C.R. Séanc. Acad. Sci. 228: 1675–1678 (sep. p. 1–4). [Œuvres IV(120)].

Grandjean, F., 1949a. Au sujet des variations individuelles et des polygones de fréquence. – C.R. Séanc. Acad. Sci. 229: 801–804 (sep. p. 1–4). [Œuvres IV(123)].

Grandjean, F., 1949b. Formules anales, gastronotiques, génitales et aggénitales du développement numérique des poils chez les Oribates. – Bull. Soc. Zool. France 74: 201–225. [Œuvres IV(125)].

Grandjean, F., 1950. Observations éthologiques sur Camisia segnis (Herm.) et Platynothrus peltifer (Koch) (Acariens). – Bull. Mus. Nat. Hist. Natur. (2) 22: 224–231, Figs. A–B. [Œuvres IV(127)].

Grandjean, F., 1950a. Etude sur les Lohmanniidae (Oribates, Acariens). – Arch. Zool. Exp. Gen. 87: 95–162, Figs. 1–9. [Œuvres IV(128)].

Grandjean, F., 1951. Les deux sortes de temps du biologiste. – C.R. Séanc. Acad. Sci. 233: 336–339 (sep. p. 1–4). [Œuvres IV(135)].

Grandjean, F., 1951a. Les relations chronologiques entre ontogenèses et phylogenèses d'après les petits caractères discontinus des Acariens. – Bull. Biol. France Belg. 85: 269–292, Figs. 1–2. [Œuvres IV(136)].

Grandjean, F., 1952. Sur les variations individuelles. Vertitions (écarts) et anomalies. – C.R. Séanc. Acad. Sci. 235: 640–643 (sep. p. 1–4). [Œuvres IV(143)].

Grandjean, F., 1952a. Observations sur les Palaeacaroïdes (2e série). – Bull. Mus. Nat. Hist. Natur. (2) 24: 460–467, Figs. 1–3. [Œuvres IV(145)].

Grandjean, F., 1954. Etude sur les Palaeacaroïdes (Acariens, Oribates). – Mém. Mus. Nat. Hist. Natur. (N.S.), sér. A, Zool. 7: 179–274, Figs. 1–25. [Œuvres IV(156)].

Grandjean, F., 1954a. Les deux sortes de temps et l'évolution. – Bull. Biol. France Belg. 88: 413–434, Figs. 1–19. [Œuvres V(160)].

Grandjean, F., 1955. Sur un Acarien des îles Kerguélen. Podacarus Auberti (Oribate). – Mém.

Mus. Nat. Hist. Natur. (N.S.), sér. A, Zool. 8: 109–150, Figs. 1–10. [Œuvres V(165)].

Grandjean, F., 1957. L'évolution selon l'âge. – Arch. Sci. Genève 10: 477–526, Figs. 1–3. [Œuvres V(184)].

Grandjean, F., 1958. Sur le comportement et la notation des poils accessoires postérieurs aux tarses des Nothroïdes et d'autres Acariens. – Arch. Zool. Exp. Gén. 96: 277–308, Figs. 1–6. [Œuvres V(190)].

Grandjean, F., 1961. Nouvelles observations sur les Oribates (1re série). – Acarologia 3: 206–231. [Œuvres VI(204)].

Grandjean, F., 1961a. Les Amerobelbidae (Oribates) (1re partie). – Acarologia 3: 303–343, Figs. 1–13. [Œuvres VI(205)].

Grandjean, F., 1961b. Considérations numériques sur les poils génitaux des Oribates. – Acarologia 3: 620–636. [Œuvres VI(207)].

Grandjean, F., 1961c. Un bel exemple d'évolution collective et vertitionnelle: la régression numérique des ongles aux pattes de Nothrus silvestris Nic. (Acarien, Oribate). – Bull. Biol. France Belg. 95: 539–555. [Œuvres VI(210)].

Grandjean, F., 1963. La néotrichie du genre Tricheremaeus d'après T. nemossensis n. sp. (Oribate). – Acarologia 5: 407–437, Figs. 1–5. [Œuvres VI(216)].

Grandjean, F., 1964. La solénidiotaxie des Oribates. – Acarologia 6: 529–556. [Œuvres VII(220)].

Grandjean, F., 1965. Nouvelles observations sur les Oribates (4e série). – Acarologia 7: 91–112, Figs. 1–2. [Œuvres VII(222)].

Grandjean, F., 1965a. Complément à mon travail de 1953 sur la classification des Oribates. – Acarologia 7: 713–734, Figs. 1–4. [Œuvres VII(225)].

Grandjean, F., 1971. Nouvelles observations sur les Oribates (8e série). – Acarologia 12: 849–876, Figs. 1–6. [Œuvres VII(238)].

Grandjean, F., 1971a. Caractères anormaux et vertitionnels rencontrés dans des clones de Platynothrus peltifer (Koch). Première partie. – Acarologia 13: 209–237 Figs. 1–8. [Œuvres VII(239)].

Grandjean, F., 1973. Caractères anormaux et vertitionnels rencontrés dans des clones de Platynothrus peltifer (Koch). Chapitres I à VI de la deuxième partie. – Acarologia 14: 454–478, 4 tab. [Œuvres VII(240)].

Grandjean, F., 1974. Caractères anormaux et vertitionnels rencontrés dans des clones de Platynothrus peltifer (Koch). Chapitres VII à XIII de la deuxième partie. – Acarologia 15: 759–780, 1 tab., 1 fig. [Œuvres VII(241)].

Hammen, L. van der, 1962. Evolutie en tijd. – Vakblad voor Biologen 42: 1–12, Figs. 1–10.

Hammen, L. van der, 1963. The addition of segments during the postembryonic ontogenesis of the Actinotrichida (Acarida) and its importance for the recognition of the primary subdivision of the body and the original segmentation. – Acarologia 5: 443–454.

Hammen, L. van der, 1964. The relation between phylogeny and post-embryonic ontogeny in Actinotrichid mites. – Acarologia 6 (fasc. h.s.): 85–90, Figs. 1–8.

Hammen, L. van der, 1970. La segmentation primitive des Acariens. – Acarologia 12: 3–10, Tab. 1, Fig. 1.

Hammen, L. van der, 1972. A revised classification of the mites (Arachnidea, Acarida) with diagnosis, a key, and notes on phylogeny. – Zool. Meded. Leiden 47: 273–292, Fig. 1.

Hammen, L. van der, 1974. L'évolution du cycle vital chez les Acariens et les autres groupes d'Arachnides. – Acarologia 15: 384–390.

Hammen, L. van der, 1975. L'évolution des Acariens, et les modèles de l'évolution des Arachnides. – Acarologia 16: 377–381, Figs. 1–2.

Hammen, L. van der, 1977. A new classification of Chelicerata. – Zool. Meded. Leiden 51: 307–319, Fig. 1, Tab. 1–3.

Hammen, L. van der, 1977a. The evolution of the coxa in mites and other groups of Chelicerata. – Acarologia 19: 12–19, Figs. 1–2.

Hammen, L. van der, 1978. The evolution of the chelicerate life-cycle. – Acta Biotheoretica 27: 44–60, Figs. 1–6. [Cf. essay III].

Hammen, L. van der, 1979. Evolution in mites, and the patterns of evolution in Arachnidea. − Proc. 4th Int. Congr. Acarol. Saalfelden 1974: 425−430, Figs. 1−2.

Hammen, L. van der, 1979a. Comparative studies in Chelicerata I. The Cryptognomae (Ricinulei, Architarbi and Anactinotrichida). − Zool. Verh. Leiden 174: 1−62, Figs. 1−31.

Hammen, L. van der, 1980. Glossary of acarological terminology 1. General terminology. − The Hague (Dr W. Junk Publishers): viii + 244 p.

Hammen, L. van der, 1981. Type-concept, higher classification and evolution. − Acta Biotheoretica 30: 3−48, Figs. 1−7. [Cf. essay V].

Hammen, L. van der, 1981a. Numerical changes and evolution in Actinotrichid mites (Chelicerata). − Zool. Verh. Leiden 182: 1−47, Figs. 1−13.

Hammen, L. van der, 1982. Comparative studies in Chelicerata II. Epimerata (Palpigradi and Actinotrichida). − Zool. Verh. Leiden 196: 1−70, Figs. 1−31.

Hammen, L. van der, 1986. Acarological and arachnological notes. − Zool. Meded. Leiden 60: 217−230, Figs. 1−3.

Hammen, L. van der, 1986a. On some aspects of parallel evolution in Chelicerata. − Acta Biotheoretica 35: 15−37, Figs. 1−9. [Cf. essay IV].

Kauffman, S.A., 1973. Control circuits for determination and transdetermination. − Science 181: 310−318, Figs. 1−7, Tab. 1−5.

Kauffman, S.A., 1987. Developmental logic and its evolution. − BioEssays 6(2): 82−87, Figs. 1−3.

Knülle, W., 1957. Morphologische und entwicklungsgeschichtliche Untersuchungen zum phylogenetischen System der Acari: Acariformes Zachv. I. Oribatei: Malaconothridae. − Mitt. Zool. Mus. Berl. 33: 97−213, Figs. 1−41.

Lions, J.-C., 1964. La variation du nombre des ongles des pattes de Rhysotritia ardua (C.L. Koch) 1836 (Acarien, Oribate). − Rev. Ecol. Biol. Sol. 1: 41−65, Tab. 1−16.

Lions, J.-C., 1967. La prélarve de Rhysotritia ardua (C.L. Koch) 1836 (Acarien, Oribate). − Acarologia 9: 273−283, Figs. 1−3.

Matsakis, J.Th., 1967. Sur certaines variations quantitatives affectant les appendices arthropodiens: cas des griffes de l'Oribate Rhysotritia ardua. − Bull. Soc. Hist. Natur. Toulouse 103: 590−631, Tab. 1−9, Figs. 1−3.

Pijnacker, L.P. & M.A. Ferwerda, 1972. Diffuse kinetochores in the chromosomes of the arrhenotokous spider mite Tetranychus urticae Koch. − Experientia 28: 354, Figs. 1−5.

Sellnick, M. & K.-H. Forsslund, 1955. Die Camisiidae Schwedens (Acar. Oribat.). − Arch. Zool. (2) 8: 473−530, Figs. 1−47.

Travé, J., 1973. The specific stability and the chaetotactic variations in Mucronothrus nasalis Willm. (Oribatei). − Proc. 3rd Int. Congr. Acarol. Prague 1971: 303−309, Tab. 1−2, Figs. 1−2.

Travé, J., 1974. Les variations chaetotaxiques dans quelques populations de Mucronothrus nasalis Willm. (Oribate). − Acarologia 15: 521−533, Tab. 1−6, Figs. 1−3.

Travé, J., 1977. La néotrichie épimérique d'Hermannia jesti sp. n. (Oribate). − Acarologia 19: 123−131, Figs. 1−12.

Travé, J., 1979. Neotrichy in Oribatid mites. − Recent Advances in Acarology 2: 523−527.

Travé, J., 1979a. La néotrichie chez les Oribates (Acariens). − Acarologia 20: 590−602.

III. THE EVOLUTION OF THE CHELICERATE LIFE-CYCLE

1. Introduction

In the previous essay, numerical changes in the individual occurrence of small elements (setae, setiform organs, segments) and in spatial patterns of these elements were dealt with. The present essay deals with complete individuals, viz., the successive forms of the chelicerate life-cycle, and the evolutionary changes in the shapes and numbers of these forms (i.e. the numerical patterns of these forms in time). It is based on a paper which I published nine years ago (Van der Hammen, 1978; it included a summary of three previous papers); the text is now entirely revised and partly rewritten, and includes data from two papers which were published afterwards (Van der Hammen, 1985: 397–399, Fig. 3; 1986: 223–224). The glossary, which was originally added to the paper, is omitted; definitions of the terms (and references to the literature) can now be found in the general volume of the *Glossary of Acarological Terminology* (Van der Hammen, 1980). It may be repeated here that, although the essay deals with complete forms, these forms are composed of segments (compartments) representing spatial variations of a single genetic pattern, and that each of the segments has, moreover, the potentiality of repeated variation in the course of ontogenetic time.

Although the postembryonic development has been studied in many groups of Chelicerata, comparative data on this development are rare, and our general knowledge of the chelicerate life-cycle is not homogeneous. This is partly caused by the great biological differences existing between the various groups. In some groups, such as Actinotrichid and Anactinotrichid mites, immature and adult forms can often be found throughout the year, and even together in one sample, whilst the various immature forms present distinct morphological differences. In many other groups of Chelicerata, however, the life-cycle can present seasonal aspects; in addition, species have often to be reared in order to study the complete development. The morphological differences between the various forms of a life-cycle can, moreover, be rather indistinct. In some cryptozoic groups (Palpigradi, Ricinulei) the life-cycle is even incompletely known.

Until recently there was also no uniform theoretical base for the general study of the postembryonic development in Chelicerata: there were no general concepts for the various types of forms, and there was no general terminology based on comparative studies.

The aim of the present essay is to lay the theoretical foundations for a comparative study of the chelicerate life-cycle and its evolution. This study was started several years ago (Van der Hammen, 1974, 1975, 1979), when I tried to apply the terminology developed in acarology to other groups of Chelicerata. It appeared that acarological terminology was not sufficient for this purpose and had to be completed, whilst some acarological concepts had to be adapted.

The present essay is a further development of my previous studies in this field. It is founded on a comparison of my own acarological knowledge with data from literature on other groups of Chelicerata. It is devoted in particular to moults and the various forms between, and to the evolutionary phenomena connected with the life-cycle.

2. Forms and moults

A comparative study of the life-cycle begins with chelicerate moulting and the forms which are the result of it (types of moulting are not discussed in the present section; see Van der Hammen, 1980: 42, *sub* dehiscence). In Mites there generally are, in each group, a fixed number of moults, and consequently a fixed number of (distinctly recognizable) instars. In most groups of Actinotrichid mites and in Opilioacarids (i.e. primitive Anactinotrichid mites) the number of instars is six (Figs. 3.1, 3.2). The same number is apparently also present in Uropygi (Weygoldt, 1971a, 1972), Schizomida (Rowland, 1972, 1972a) and Pseudoscorpionida (Weygoldt, 1968, 1969, 1971). Five instars are now known from Ricinulei (Pittard & Mitchell, 1972), but the possible occurrence of a regressive prelarva, which could be present inside the egg, has not yet been studied in this group.

There are, however, also groups with a less constant number of moults. In Scorpions, there are apparently 6−8 moults (Aubert, 1959, 1963; Vachon, 1952; Vachon, Roy & Condamin, 1970). Numerous moults are mentioned in the case of Amblypygi (Weygoldt, 1970), where the adult continues moulting. There is also a variable number of moults in Spiders (Holm, 1941; Vachon, 1958, 1959, 1965; Legendre, 1965; Emerit, 1969, 1972), where in some groups the adults also continue moulting. A rather great and variable number of moults is mentioned for Opilionida and Solifugae (Juberthie, 1964, 1965; Muma, 1966).

Not every moult has the same effect. In many cases moulting results in a form distinctly different from the preceding form. Sometimes, however, moulting brings no change, or there is only a difference in size; in these cases the moults are called repetition-moults and growing-moults (Grandjean, 1970). Growing-moults are, for instance, present in the case of moulting adults of Amblypygi and Spiders.

Among the forms which are the result of moulting, three types can now be distinguished. Forms which are the result of growing-moults or repetition-moults (type 1) are called isophena (Van der Hammen, 1975). The forms representing types 2 and 3 are characterized by distinct, discontinuous differences from the form immediately preceding them; they are called instars. Type 2 is found in life-cycles of groups with an equal number of moults; in this case the levels of development of species of these groups can be homologized. Type 3 is found in groups in which the life-cycle presents a variable number of moults, and in which homologous levels cannot be distinguished. Recently (Van der Hammen, 1975), I have restricted the stase concept, introduced by Grandjean (1938), to the case of homologizable levels (type 2). Forms of type 3 are now called stasoids (Van der Hammen, 1975). In one life-cycle, stases and stasoids can both be present, such as in Scorpions where the first instar certainly is a stase, whilst the other immatures are apparently all stasoids.

3. Phases

In a sequence of stases the instars can be more or less similar, or present important differences from the preceding and succeeding stases. In the first case they are called homostases, in the second case they are called heterostases (Grandjean, 1954). In many Mites we can distinguish a prelarva, a larva, three more of less similar nymphs, and an adult. In this case the three nymphs are homostases. The life-cycle can be subdivided into phases (Vachon, 1953), according to the presence of homostases and heterostases. In the case of Mites we can distinguish four phases: a prelarval, a larval, a nymphal and an adult phase (Figs. 3.1, 3.2). In Chelicerata the number of phases varies. Generally there are three or four phases. Theoretically, a life-cycle with two phases (an immature and an adult phase) is also possible; this type of life-cycle could have been present in the ancestors of extant Chelicerata.

The phase-concept could also be applied to series of stasoids. The concept enables us to standardize the terminology of the subsequent forms of a life-cycle. Until now there have been two criteria to define prelarva, larva, prenymph and nymph: the criterion of the level of development and the criterion of regression. Acarologists have used the first criterion. In Mites (cf. Van der Hammen, 1976, 1980) the instar of level 1 is called prelarva, the instar of level 2 larva, etc. Araneologists, on the contrary, have used the criterion of regressive evolution. According to the degree of regression one or two prelarvae have generally been distinguished, and one or two larvae. Application of the phase concept to the terminology of instars of various levels, enables us to standardize the names of chelicerate instars. In the (theoretical) case of two phases, the instars are called nymphs and adult; in the case of three phases we can also distinguish larvae; in the case of four phases we can moreover distinguish prelarvae. In this terminology the term prenymph could only be used in the case of an additional (fifth) phase between larval and nymphal phase.

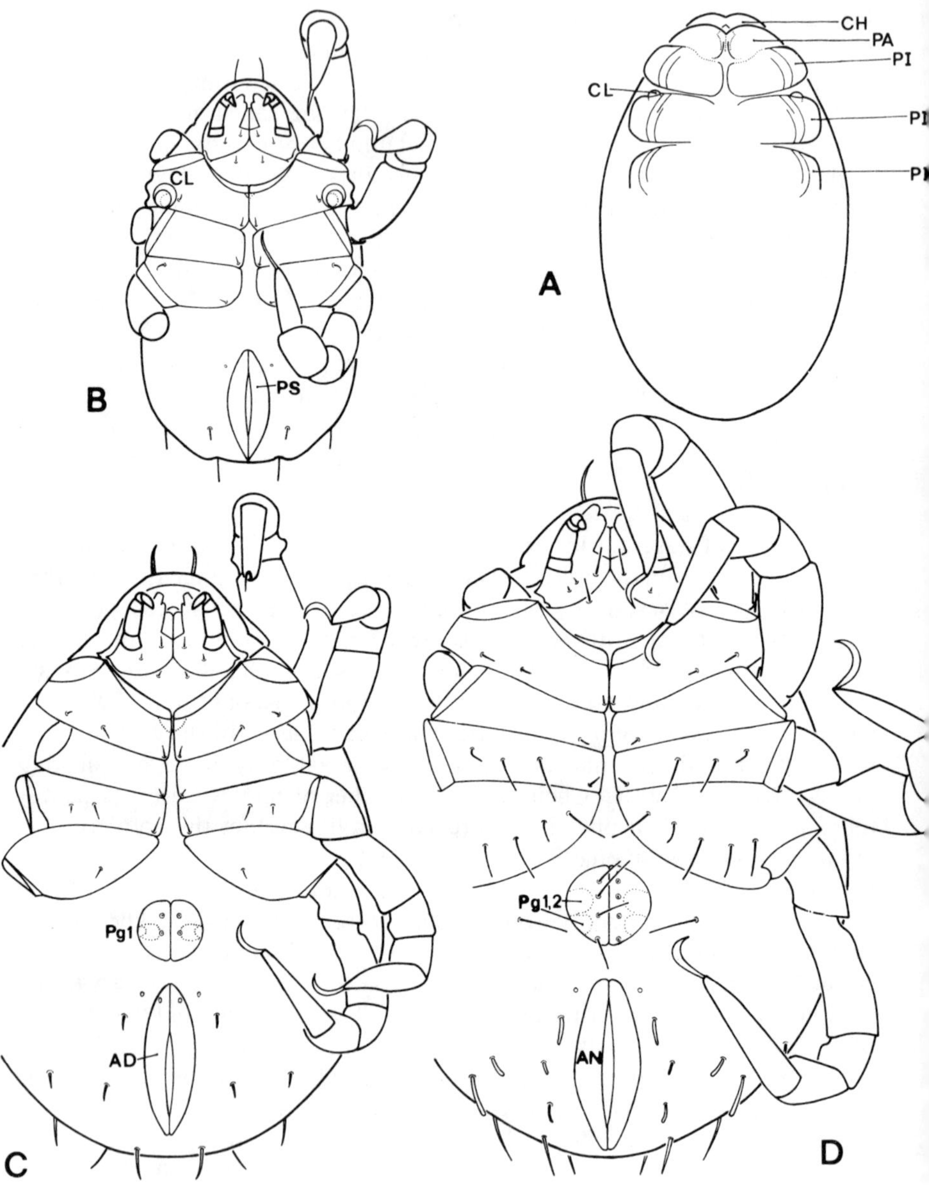

Fig. 3.1. Postembryonic development of *Hermannia convexa* C.L. Koch (Actinotrichida: Oribatida) (1); ventral views of: A: prelarva; B: larva; C: protonymph; D: deutonymph; *AD*: adanal segment; *AN*: anal segment; *CH*: chelicera; *CL*: Claparède's organ; *PA*: palp; *PI-III*: legs I-III; *PG1–2*: genital papillae; *PS*: pseudanal segment; *A–D*: × 118. (After Van der Hammen, 1972).

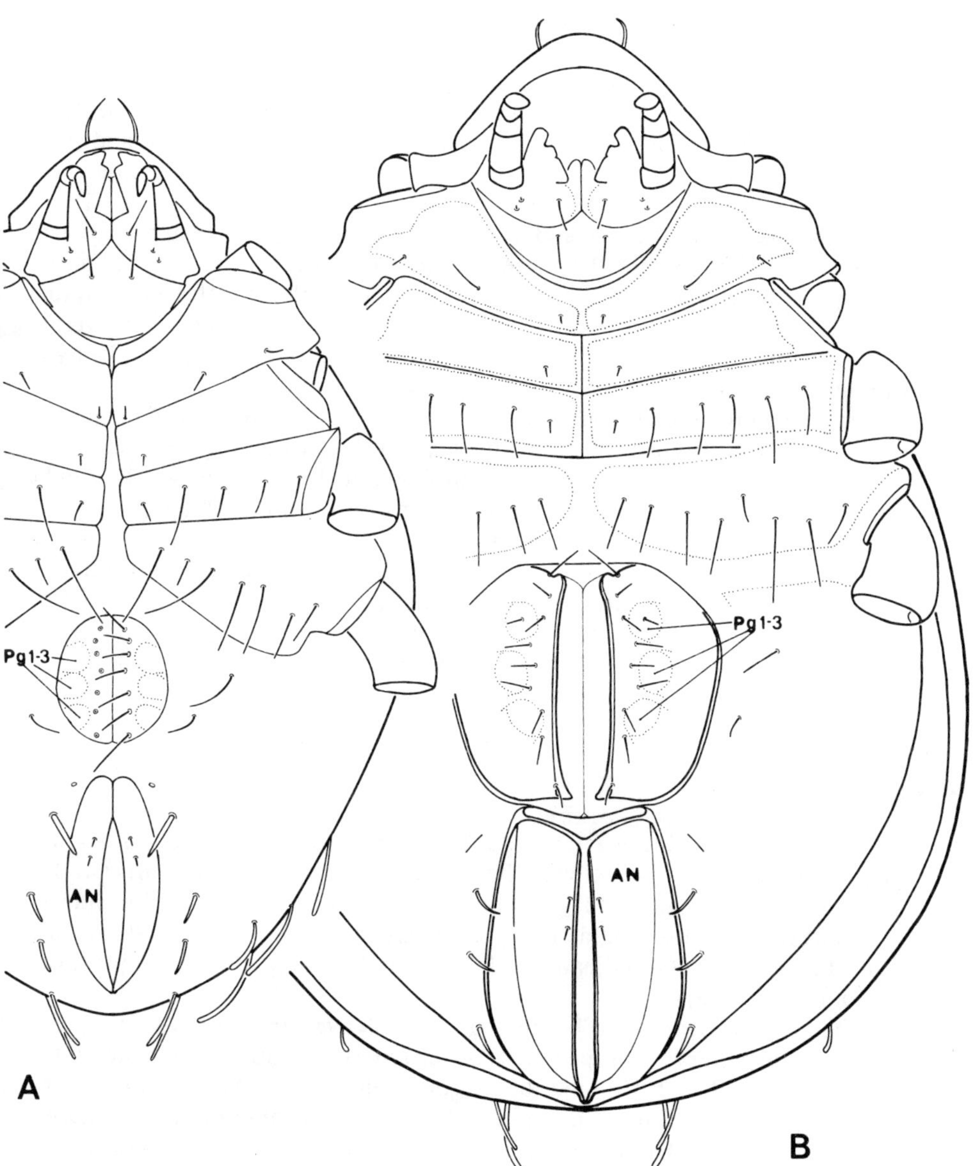

Fig. 3.2. Postembryonic development of *Hermannia convexa* C.L. Koch (Actinotrichida: Oribatida) (2); ventral views of: A: tritonymph; B: adult female; *AN*: anal segment; *PG1–3*: genital papillae; A–B: × 118. (After Van der Hammen, 1972).

In order to standardize terminology also in each separate group, the number of phases on which the terminology in a group is based, should be that of the primitive condition of that group. In Actinotrichid mites, where generally three nymphs can be distinguished, constituting one nymphal phase, the three nymphs of Trombidina (Actinedida) are still named proto-, deuto- and tritonymph, although in reality they represent heterostases constituting three separate phases.

In the case of stasoids a standardization, comparable to that in groups with stases, will be impossible; the phase concept will be here the main base of terminology. The terminology of stasoids should preferably be different from that of stases. Stasoids of the nymphal phase could, for instance, be called nymphoids; the nymphal phase of Spiders, e.g., apparently comprizes 5–10 nymphoids.

Isophena preserve, of course, the name of the level of which they form part. We could, for instance, speak of the isophena of nympha 2.

As examples of a standardized terminology, the following life-cycles can be mentioned. In Scorpions there are apparently a larva, several nymphoids and an adult. In Amblypygi there are apparently a larva, several nymphoids, an adult and isophena of the adult. In Uropygi there are a larva, four nymphs and an adult. in Xiphosura there are four phases (a prelarval, a larval, a nymphal and an adult phase); the prelarval phase remains inside the egg (there are two prelarval moults); the larval phase comprizes the so-called trilobitic larva; the nymphal phase comprizes the so-called prestwickianella larvae.

4. Regressive forms

In the course of evolution, stases (it is not known from stasoids) can be subject to various regressive evolutions. They can loose the use of mouthparts (so that they are unable to eat), or they can loose the use of mouthparts and appendages (so that they are also unable to walk). Regressive stases of the first type are called elattostases (Grandjean, 1957a), those of the second type calyptostases (Grandjean, 1938).

The larva of Scorpions, Uropygi and Amblypygi, the larva (or larvae) of Spiders, the prelarva of a species of *Saxidromus* (Actinotrichida: Actinedidia) (cf. Coineau, 1974, 1979), and the deutonymph of many Acaridida (= Acaridiae, a group of Actinotrichid mites) (Fig. 3.3) are examples of elattostases.

The prelarva of Opilioacarida is a calyptostase, although it constitutes a transition to an elattostase (Coineau, 1973) (Fig. 3.4B).

The prelarva(e) of Spiders, the larvae of Pseudoscorpionida, the prelarva of many Mites (cf. Travé, 1976) (Figs. 3.1A, 3.4A) and the prelarva, the proto- and the tritonymph of Trombidina (Actinotrichida) (Fig. 3.5) (cf. Grandjean, 1957, 1959) are examples of calyptostases. The examples from Actinotrichida demonstrate that a calyptostase can take the place of an elattostase at the same homologous level.

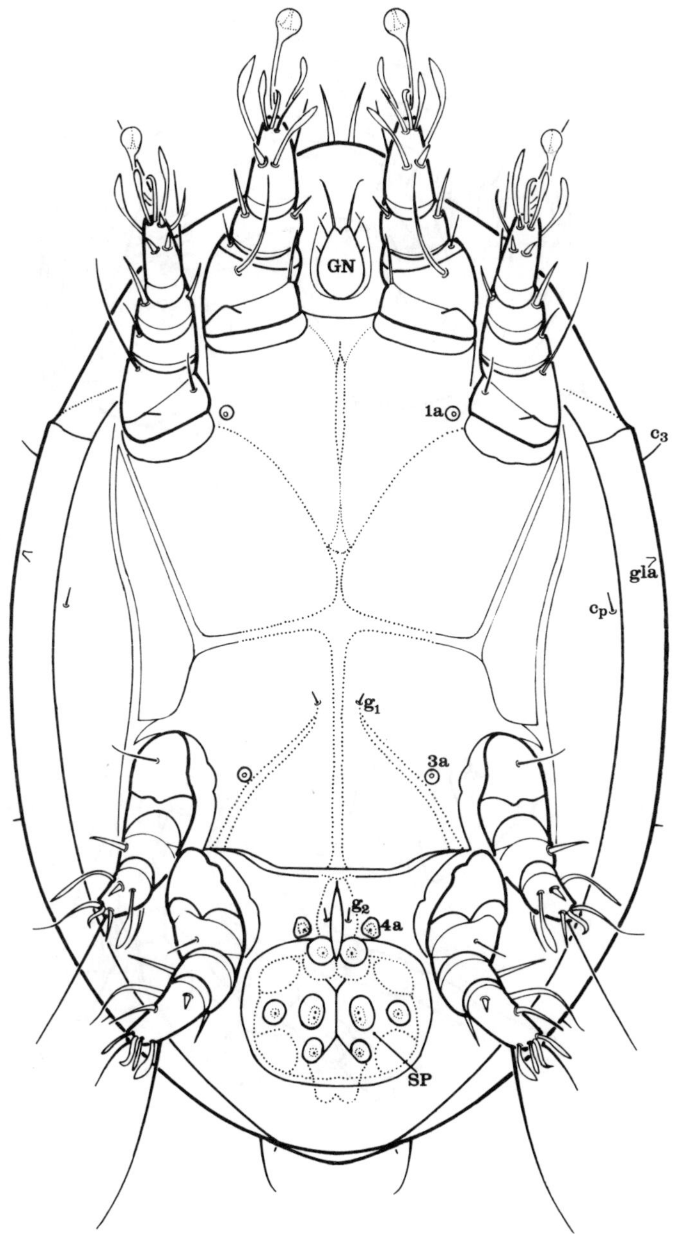

Fig. 3.3. Rhizoglyphus echinopus (Fumoure & Robin) (Actinotrichida: Acaridida), ventral view of hypopus (elattostasic deutonymph); c_3, c_p: notogastral setae; g_{1-2}: genital setae; *gla*: latero-opisthosomatic gland; *GN*: regressive gnathosoma; *SP*: sucker plate; *1a, 3a, 4a*: epimeral setae; × 368. (After Van der Hammen, 1972).

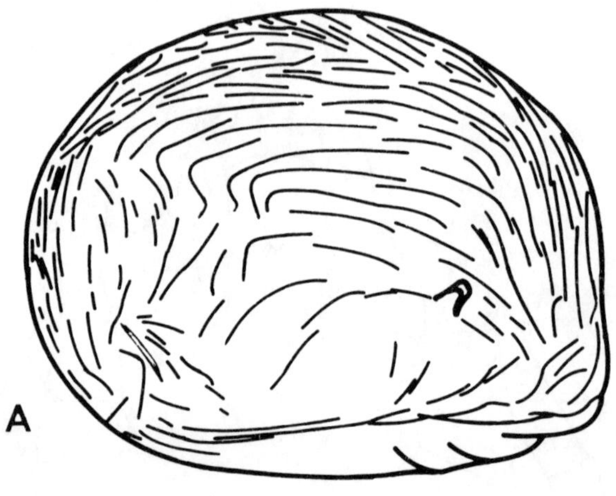

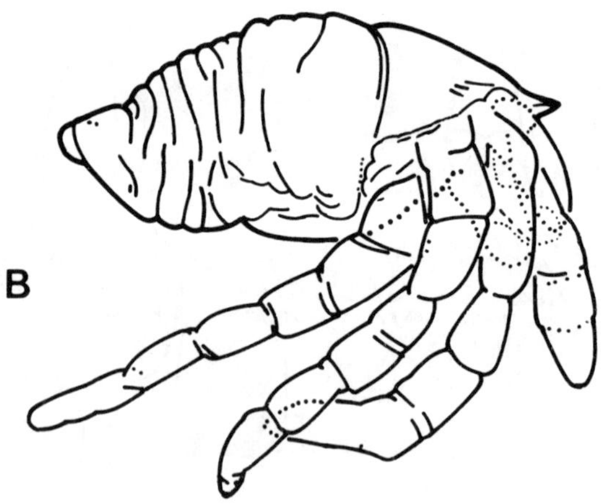

Fig. 3.4. Prelarvae of two species of mites, in lateral orientation; A: *Hoplophthiracarus* spec., B: *Phalangiacarus brosseti* Coineau & Van der Hammen; A: × 488; B: × 113. (Fig. 3.4A after Van der Hammen, 1963; Fig. 3.4B redrawn after a drawing by Coineau in Van der Hammen, 1976).

5. Evolutionary phenomena

The chelicerate life-cycle can be subject to various types of evolution. A very characteristic type of evolution is the regression of the first instar(s) (Fig. 3.1 A, B, 3.4, 3.5A); as far as known, this regression is present in all groups of extant Chelicerata. The first instar of Scorpions, Amblypygi, Uropygi and Opilionida is an elattostase or a calyptostase. In Spiders several of the first instars can be subject to regression. In Actinotrichid mites the first stase generally is a calyptostase, but in one primitive group it can be an elattostase (Coineau, 1974, 1979). In Opilioacarida (Fig. 3.4 B) it is a calyptostase with some characters of an elattostase (Coineau, 1973; Coineau & Van der Hammen, 1979). The larva of Mites (Fig. 3.1 B) and Ricinulei has only three pairs of legs (there can be a vestige of leg IV); this is also the result of a regressive evolution. In some species of Mites the larva has even become an elattostase. The evolutionary phenomenon of the regression of the first levels of postembryonic development is called protelattosis. (Van der Hammen, 1975). It comprises the regression of separate structures (such as leg IV), the formation of elattostases, and the formation of calyptostases.

A second evolutionary phenomenon is the attainment of maturity at a previously immature level. It is generally called neoteny. Neoteny is apparently possible, or probable or even practically certain in a number of cases: for instance in the case of some male Scorpionida and Spiders, and in those Pseudoscorpionida and Mites where two nymphs are present instead of three.

A third evolutionary phenomenon pertains to the regression of instars at other levels than at the beginning of postembryonic development. It is called metelattosis (Van der Hammen, 1975). This phenomenon is well-known from Actinotrichid mites. I point out the formation of hypopi in Tyroglyphid mites (an elattostase at the deutonymphal level) (Fig. 3.3), and the formation of calyptostases at the level of nymphae 1 and 3 in Trombidina (Actinotrichida: Actinedida) (Fig. 3.5). This phenomenon is apparently very rare in other groups of Chelicerata. I can cite only one example discovered by Weygoldt (1970): the disharmonic regression of trichobothria in the third instar of Amblypygi. This case apparently represents the very first beginning of a separation of the third instar from its phase.

The fourth evolutionary phenomenon concerns the formation of isophena at a certain level of development; it is called plethomorphosis (Van der Hammen, 1975). This phenomenon is found in nymphs of Argasid ticks (Anactinotrichida: Ixodida), and in the case of moulting adults (such as in Amblypygi and some groups of Spiders). It could also be present in nymphs of several other groups, but unmistakable data are lacking.

6. General model of the evolution of the chelicerate life-cycle

We can now try to arrange the life-cycles, the various types of forms or instars,

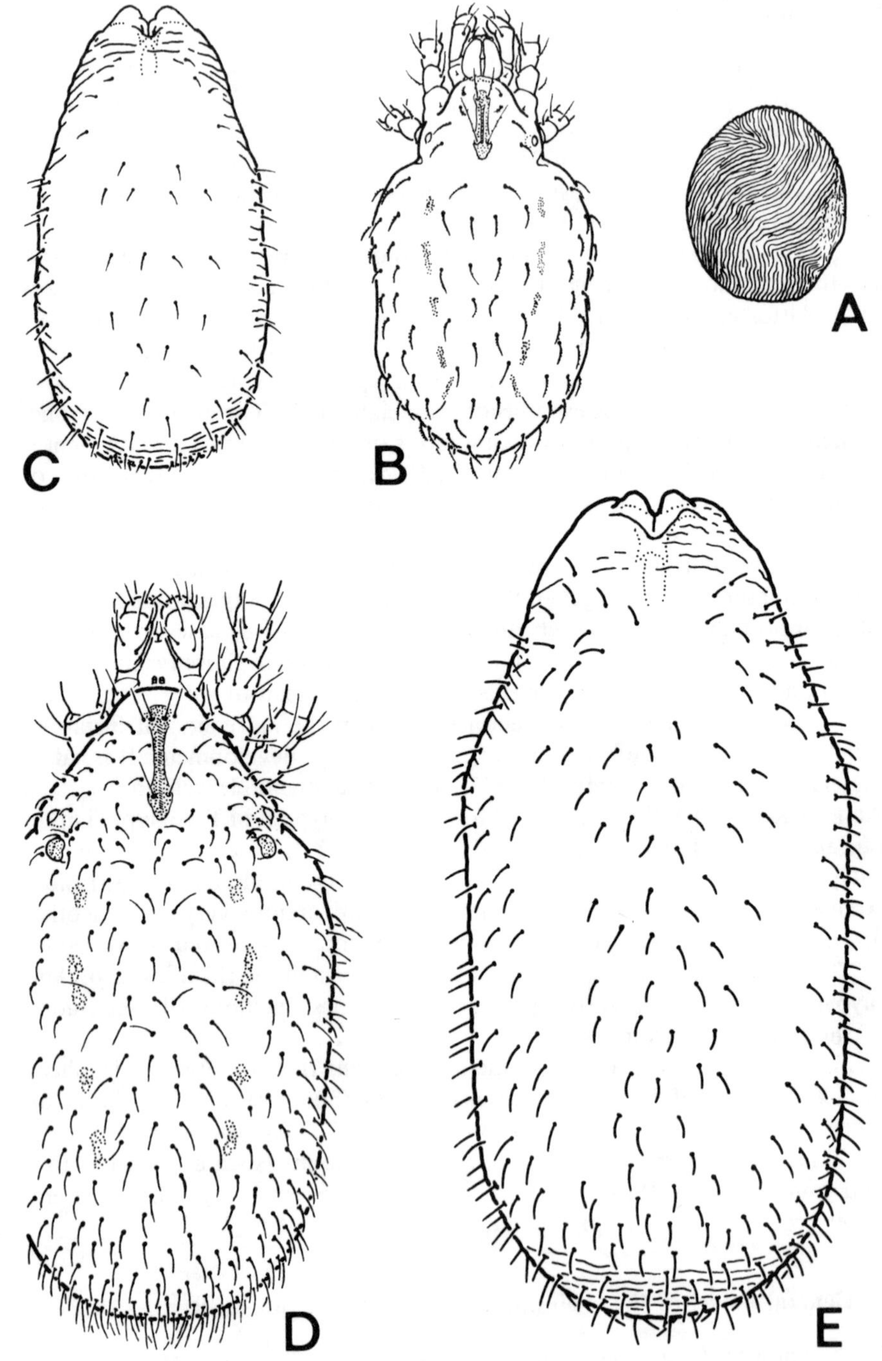

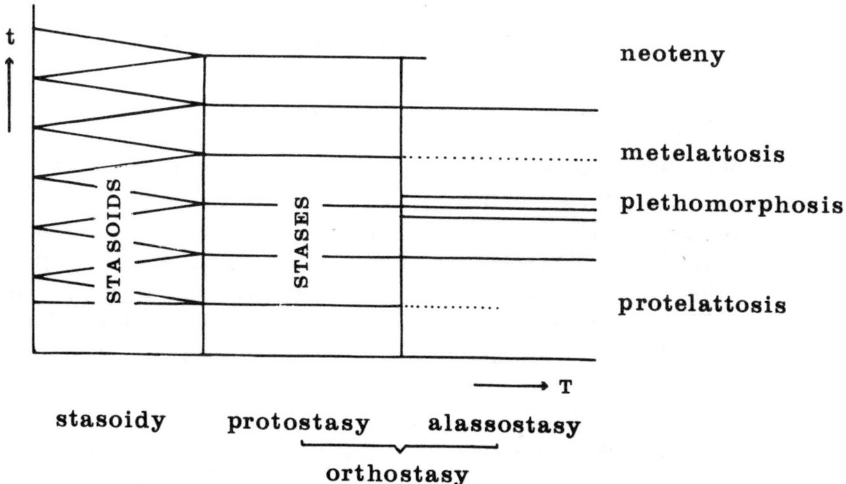

Fig. 3.6. Model (chronological diagram) of the evolution of the chelicerate life-cycle.

and the evolutionary phenomena in a chronological diagram in which the phylogenetic time is on the horizontal axis, and the ontogenetic time on the vertical axis (Fig. 3.6). I have introduced the hypothesis that the evolution of the chelicerate life-cycle has started with stasoids and presented only one immature phase; it is supposed that stases arose from stasoids. According to the presence of stasoids or stases, two stages of evolution can now be distinguished, which are called stasoidy and orthostasy (Van der Hammen 1975, 1979).

Between these two stages there is certainly a period of transition in which stases and stasoids are both present. In several groups, such as Scorpions, Amblypygi and Spiders, the nymphal phase apparently consists of stasoids, whilst the first one or two phases probably consist of stases.

It is supposed that the ancestral number of stases in orthostasy has, in all cases, been six (as still found in several groups). In the course of evolution the ancestral orthostasic life-cycle has been subject to the evolutionary phenomena already mentioned (protelattosis, plethomorphosis, metelattosis and neoteny). Consequently orthostasy can be subdivided into protostasy (the ancestral orthostasic condition) and alassostasy (the derived condition) (Van der Hammen, 1975).

The diagram constitutes a hypothetic model of the evolution of the postembryonic development in Chelicerata. It signalizes many gaps in our knowl-

---

←

Fig. 3.5. Part of the life-cycle of *Balaustrium florale* Grandjean; A: calyptostasic prelarva; B: larva; C: calyptostasic protonymph; D: deutonymph; E: calyptostasic tritonymph; A–E: × 120. (Redrawn after Grandjean, 1957, 1959).

edge, and a long program of investigations could be based on it. Simple questions could probably be easily solved; it will, for instance, not be difficult to check whether the nature and number of the forms of a life-cycle are, in some way, related to the duration of the cycle and the presence of seasonal aspects. It will be especially interesting to make a further study of the postembryonic development of Opilionida and Solifugae. In these two groups the number of moults is often great, but apparently variable; we do not know which types of forms or instars are found (in some cases stasoids, stases as well as isophena could be present in one life-cycle). It will also be very important to complete the study of the postembryonic development of Palpigradi and Ricinulei. The life-cycle of the last-mentioned group resembles that of Opilioacarida, but a calyptostasic or elattostasic prelarva has not yet been found.

The present study has led to a perfection of the terminology with reference to the life-cycle and its evolution. The hypothetic model is still very simple and does not account for the overlapping of the periods. At present it is probably sufficient, but it certainly is not definitive. We must be ready to adapt it to new observational data and new hypotheses.

It may be remarked here that the possibility cannot be excluded (it is, on the contrary, rather probable) that, as to phylogeny, the model should be read in both directions and that a third period (neostasoidy) characterized by the presence of neostasoids, and which period is still indistinguishable from stasoidy, should be recognized. As evolutionary phenomenon, the origination of neostasoids is comparable with the origination of neotrichy where single setae are replaced by fields (spatial patterns) of setae. In the case of neostasoidy, the instars no longer evolve as separate forms (i.e. by age), but as groups of variable numbers of successive forms (i.e. patterns in time). The laws of the numerical evolution of the life-cycle are apparently similar to those of the numerical evolution of setae, etc.

7. Remarks

The evolutionary phenomena, described above, can be compared with evolutionary trends found in various other animal groups. One of the most striking of these is the divergent evolution of adaptation at different levels of postembryonic development, which divergence is always associated with radical metamorphosis (tadpole and frog; caterpillar, chrysalis and butterfly; etc.). In this case, deviation from the ancestral way of life can arise either at one or more of the immature levels or at the level of the adult (in the first-mentioned case, these immatures are more advanced than the adults). In Trombidina (Actinotrichid mites), e.g., the parasitic larva has deviated from the ancestral way of life, which is still found in the predatory nymph and adult.

A second general phenomenon is constituted by the increasing dependence of the young, in the course of evolution, on the parents. In all Chelicerata, the first stase, or the first stases, are at least elattostasic (they are, moreover, often carried on the back of the mother), but can also have become calyptostasic (the last-mentioned condition can give rise to ovoviviparity). In birds, e.g., young which are nidifugous (leaving the nest after hatching) and precocious (requiring little care after hatching) are regarded as representing the ancestral condition, whilst the derived condition is regarded as being represented by young which are nidicolous (living in the nest for a time after hatching) and altricial (requiring care after hatching).

It will be interesting to consider whether some of the cases of metelattosis mentioned above (the origination of calyptostasic proto- and tritonymphs in Trombidina) represent also cases of a more general evolutionary phenomenon, viz., shortening of the life-cycle (as in the case of neoteny).

References

Auber, M., 1959. Observations sur le biotope et la biologie du Scorpion aveugle: Belisarius xambeui E. Simon. – Vie et Milieu 10: 160–167, Figs. 1–3, 1 tab.

Auber, M., 1963. Reproduction et croissance de Buthus occitanus AMX. – Ann. Sci. Nat., Zool. (12), 5: 273–286, Figs. 1–9, 1 tab.

Coineau, Y., 1973. A propos de quelques caractères particulièrement primitifs de la prélarve et de la larve d'un Opilioacaride du Gabon (Acariens). – C.R. Séanc. Ac. Sci. 276 D: 1181–1184, Figs. 1–2.

Coineau, Y., 1974. Un type nouveau d'Acariens Prostigmates libres: les Saxidromoidea nouvelle super-famille. – C.R. Acad. Sc. Paris 278 D: 1059–1062, 1 fig.

Coineau, Y., 1979. Les Adamystidae, une étonnante famille d'Acariens Prostigmates primitifs. – Proc. 4th Int. Congr. Acar. Saalfelden 1974: 431–435, Figs. 1–2.

Coineau, Y. & L. van der Hammen, 1979. The postembryonic development of Opilioacarida, with notes on new taxa and on a general model for the evolution. – Proc. 4th Int. Congr. Acar. Saalfelden 1974: 437–441, Figs. 1–2.

Emerit, M., 1969. Contribution à l'étude des Gastéracanthes (Aranéides, Argiopidés) de Madagascar et des îles voisines. – Thèse Fac. Sci. Montpellier AO-2888: 1–434 + i-xxix, Pls. 1–99.

Emerit, M., 1972. Le développement des Gastéracanthes (Araneida, Argiopidae). Une contribution à l'étude de la morphologie de l'appendice aranéidien. – Ann. Mus. Tervuren, Sci. Zool. 195: xiii + 1–103, Figs. 1–15, Tab. 1–4, Pls. 1–6.

Grandjean, F., 1938. Sur l'ontogénie des Acariens. – C.R. Séanc. Ac. Sci. 206: 146–150.

Grandjean, F., 1954. Les deux sortes de temps et l'évolution. – Bull. Biol. France Belg. 88: 413–434, Figs. 1–19.

Grandjean, F., 1957. Les stases du développement ontogénétiques chez Balaustium florale (Acarien, Erythroïde). Première partie. – Ann. Soc. Ent. France 125: 135–152, Figs. 1–5.

Grandjean, F., 1957a. L'évolution selon l'âge. – Arch. Sci., Genève 10: 477–526, Figs. 1–3.

Grandjean, F., 1959. Les stases du développement ontogénétique chez Balaustium florale (Acarien, Erythroïde). Deuxième partie. – Ann. Soc. Ent. France 128: 159–177, Figs. 1–5.

Grandjean, F., 1970. Stases. Actinopiline. Rappel de ma classification des Acariens en 3 groupes majeurs. Terminologie en soma. – Acarologia 11: 796–827, Fig. 1.

Hammen, L. van der, 1963. The Oribatid family Phthiracaridae I. Introduction and redescription of Hoplophthiracarus pavidus (Berlese). – Acarologia 5: 306–317, Figs. 1–8.

Hammen, L. van der, 1972. Spinachtingen – Arachnida IV. Mijten-Acarida. Algemene inleiding in de acarologie. – Wetenschappelijke Mededelingen Kon. Ned. Natuurhist. Ver. 91: 71 p., 37 figs.

Hammen, L. van der, 1972a. A revised classification of the mites (Arachnida, Acarida) with diagnoses, a key, and notes on phylogeny. – Zool. Meded. Leiden 47: 273–292. Fig. 1.

Hammen, L. van der, 1974. L'évolution du cycle vital chez les Acariens et les autres groupes d'Arachnides. – Acarologia 15: 384–390.

Hammen, L. van der, 1975. L'évolution des Acariens, et les modèles de l'évolution des Arachnides. – Acarologia 16: 377–381, Figs. 1–2.

Hammen, L. van der, 1976. Glossaire de la terminologie acarologique 2. Opilioacarida. – The Hague (Dr W. Junk Publishers): viii + 144 p., Tab. 1–3, Figs. 1–31, Pls. 1–5.

Hammen, L. van der, 1977. A new classification of Chelicerata. – Zool. Med. Leiden 51: 307–319, Fig. 1, Tab. 1–3.

Hammen, L. van der, 1978. The evolution of the chelicerate life-cycle. – Acta Biotheoretica 27: 44–60, Figs. 1–6.

Hammen, L. van der, 1979. Evolution in mites, and the patterns of evolution in Arachnidea. – Proc. 4th Int. Congr. Acar. Saarfelden 1974: 425–430, Figs. 1–2.

Hammen, L. van der, 1980. Glossary of acarological terminology 1. General terminology. The Hague (Dr W. Junk Publishers): viii + 144 p., Tab. 1–3. Figs. 1–31, Pls. 1–5.

Hammen, L. van der, 1985. A structuralist approach in the study of evolution and classification. – Zool. Meded. Leiden 59: 391–409, Figs. 1–5. [Cf. essay VI.]

Hammen, L. van der, 1986. Acarological and arachnological notes. – Zool. Meded. Leiden 60: 217–230. Figs. 1–3.

Holm, A., 1941. Studien über die Entwicklung und Entwicklungsbiologie der Spinnen. – Zool. Bidr. Uppsala 19: 1–214, Figs. 1–48, Pls. 1–11.

Juberthie, C., 1964. Recherches sur la biologie des Opilions. – Ann. Spéléol. 19: 5–238, Pls. 1–4. Textfigs. 1–75, Tab. 1–25.

Juberthie, C., 1965. Données sur l'écologie, le développement et la reproduction des Opilions. – Rev. Ecol. Biol. Sol 2: 377–396, Figs. 1–15, Tab. 1–2.

Legendre, R., 1965. Morphologie et développement des Chélicérates. Embryologie, développement et anatomie des Aranéides. – Fortschr. Zool. 17: 238–271, Figs. 1–16.

Muma, M.H., 1966. The life of Eremobates durangonus (Arachnida: Solpugida). – Florida Ent. 49: 233–242, Figs. 1–6, Tab. 1–2.

Pittard, K. & R.W. Mitchell, 1972. Comparative morphology of the life stages of Cryptocellus pelaezi (Arachnida, Ricinulei). – Grad. Stud. Tex. Tech. Univ. 1: 3–77, Figs. 1–130.

Rowland, J.M., 1972. The brooding habits and early development of Trithyreus pentapeltis (Cook), (Arachnida: Schizomida). – Ent. News 83: 69–74, Figs. 1–2.

Rowland, J.M., 1972a. Origins and distribution of two species groups of Schizomida (Arachnida). – Southwest. Natural. 17: 153–160, Figs. 1–2.

Travé, J., 1976. Les prélarves d'Acariens. Mise au point et données récentes. – Rev. Ecol. Biol. Sol. 13: 161–171.

Vachon, M., 1952. Scorpions collectés au Maroc par M.M.P. Strinati et V. Aellen (Mission scientifique Suisse au Maroc, Août-Septembre 1950). – Bull. Mus. Nat. Hist. Natur. (2) 23: 621–623.

Vachon, M., 1953. Commentaires à propos de la distinction des stades et des phases du développement post-embryonnaire chez les Araignées. – Bull. Mus. Nat. Hist. Natur. (2) 25: 294–297.

Vachon, M., 1958. Contribution à l'étude du développement post-embryonnaire des Araignées. Première note. Généralités et nomenclature des stades. – Bull. Soc. Zool. France 82: 337–354, Figs. 1–11.

Vachon, M., 1959. Contribution à l'étude du développement post-embryonnaire des Araignées. Deuxième note. Orthognathes. – Bull. Soc. Zool. France 83: 429–461, Figs. 12–63.

Vachon, M., 1965. Contribution à l'étude du développement post-embryonnaire des Araignées. Troisième note. Pholcus phalangioides (Fussl.) (Pholcidae). – Bull. Soc. Zool. France 90: 607–620, Figs. 64–73.

Vachon, M., R. Roy & M. Condamin, 1970. Le développement post-embryonnaire du Scorpion Pandinus gambiensis Pocock. – Bull. Inst. Franç. Afr. Noire 32A: 412–432, Figs. 1–14, Tab. A.

Weygoldt, P., 1968. Vergleichend-embryologische Untersuchungen an Pseudoscorpionen IV. Die Entwicklung von Chthonius tetrachelatus Preyssl., Chthonius ischnocheles Hermann (Chthoniinea, Chthoniidae) and Verrucaditha spinosa Banks (Chthoniinea, Tridenchthoniidae). – Zeitschr. Morph. Tiere 63: 111–154, Figs. 1–31.

Weygoldt, P., 1969. The biology of Pseudoscorpions. – Cambridge, Massachusetts: xiv + 145, 114 figs.

Weygoldt, P., 1970. Lebenszyklus und postembryonale Entwicklung der Geisselspinne Tarantula marginemaculata C.L. Koch (Chelicerata, Amblypygi) im Laboratorium. – Zeitschr. Morph. Tiere 67: 58–85, Figs. 1–15, Tab. 1–4.

Weygoldt, P., 1971. Vergleichend-embryologische Untersuchungen an Pseudoscorpionen V. Das Embryonalstadium mit seinem Pumporgan bei verschiedenen Arten und sein Wert als taxonomisches Merkmal. – Zeitschr. Zool. Syst. Evolutionsforsch. 9: 3–29, Figs. 1 – 36.

Weygoldt, P., 1971a. Notes on the life history and reproductive biology of the Giant Whip Scorpion, Mastigoproctus giganteus (Uropygi, Thelyphonidae) from Florida. – Journ. Zool., London 164: 137–147, Figs. 1–6, Tab. 1, Pls. 1–3.

Weygoldt, P., 1972. Geisselskorpione und Geisselspinnen (Uropygi und Amblypygi). – Zeitschr. Köln. Zoo 15: 95–107, Figs. 1–18.

IV. ON SOME ASPECTS OF PARALLEL EVOLUTION IN CHELICERATA

1. Introduction

The present essay is based on the text of a paper recently published by me in Acta Biotheoretica (Van der Hammen, 1986). It is now entirely revised and partly rewritten, particularly because, after the completion of my comparative studies in Chelicerata (Van der Hammen, 1986a), I arrived at a deeper understanding of chelicerate relationships and of the evolution of the chelicerate appendages and mouthparts.

Parallel evolution (or parallelism) is generally defined (cf. Simpson, 1961: 78, 103–106; Mayr, 1969: 82, 202, 243) as the separate development of similar characters (or the separate manifestation of similar changes in character states) in two or more lineages of common ancestry. The phenomenon (it is of fundamental importance to phylogenetic classification) is generally attributed to a common evolutionary potentiality, and to a co-ordination of these internal forces with the external forces of the environment. Parallelism must be clearly distinguished from convergence, which concept pertains to the separate development of similar characters in two or more lineages without common ancestry (cf. Simpson, 1961: 78, 103–105; Mayr, 1969: 202, 226–228). The two concepts are, theoretically, closely related to the concepts of homology and analogy: parallel evolution pertains to homologous, convergence to analogous characters. Gosliner & Ghiselin (1984: 258) recently demonstrated that, in practice, parallelism and convergence are not always easy to differentiate.

The concept of homology (Remane, 1956: 28–93) can be defined as similarity in form and topography of a structural element in different organisms (specimens of one species, representatives of different species or of supraspecific taxa) and, by extension, as similarity in any topological relations; in this way the concept can also be applied to similar changes in the character states of identical structural elements. There are several criteria for the homology of an element in different organisms, among which identical position between other structural elements, identical special characters of the element in different organisms, and the presence of transitions (in the case of differences in form of the same element in different organisms). Homology cannot be defined in terms of phylogeny (this would include circular reasoning; phylogenetic speculation can, however, be used as a hypothetical explanation of homology), and also not in terms of ontogeny (there are many examples of structural elements with different ontogeny, which are undoubtedly homologous). Because the problem of homology cannot always be solved with absolute certainty, a quantitative concept has been developed, which permits the distinction of degrees of homology (Sneath & Sokal, 1973: 77–78); this quantitative concept pertains to the degree of conformity to the above-mentioned criteria.

The concept of analogy (Simpson, 1961: 79; Mayr, 1969: 85) can be defined as similarity in function (accompanied by similarity in form), in different organisms, of structural elements which are not identical (which do not conform to the criteria of homology).

Undoubted examples of convergence are constituted by the occurrence of wings in both insects and birds (in the case of the wings of birds and bats, it is already difficult to attribute the similarities to convergence, as part of the elements of the wings are homologous), and the occurrence of legs in both tetrapod Vertebrata and Arthropoda; there are, moreover, many examples of convergence in general shape (such as in the case of whale and fish).

The concept of parallel evolution is initially formed (as in the case of the concept of homology) in an empirical way, and an unequivocal definition is hardly attainable. The recognition of parallel evolution depends, in fact, on a judgement that is based on criteria, of which the following are mentioned here. Because parallel evolution pertains to homologous characters, the criteria first applied are those of homology. Thereupon, the most important criterion of parallel evolution of a character is constituted by its isolated occurrence in two or more related taxa, separated by discontinuities, and differentiated by attributes which, after a profound analysis (including phyletic weighting), are considered taxonomically more important than the character in question (provided that the character in question is not a common character of a taxon at a higher level, in which these taxa are placed). Assumptions on parallelism can proceed from the comparison of plan of construction and phylogenetic diagram, whilst in the explanation the hypothesis of parsimony (Camin & Sokal, 1965) can play a certain part. In the following sections, these views are further explained.

In the present essay some aspects of parallel evolution in the subphylum Chelicerata are studied; in this group the recognition of the phenomenon of parallelism is of considerable importance to higher classification. Particular attention is paid to the role of changes in systems of interactions in parallel evolution, to the special case of convergence as a result of heterologous regulatory mechanisms, to parallel evolution in homonomous structures (and the superposition of parallelisms and divergencies), and to parallelisms in the evolution of the most important characters used in the classification of taxa at the superclass, class, subclass and superorder level (characters pertaining to locomotion and ingestion). At the end of the essay a few general conclusions are drawn.

2. Parallel evolution and convergence in the trichobothridial regression of Oribatid mites

Various aspects of the problem of parallel evolution can easily be demonstrated in the case of the so-called trichobothridial regression in Oribatid mites. A trichobothrium is a compound structure (a vibro- and anemoreceptor), consisting of a small cavity (the bothridium) and a variously shaped seta (bothridial seta or sensillus); it is known from many groups of Chelicerata. In Oribatid mites, a pair of trichobothria is usually present on the prodorsum (the dorsal region of the prosoma). In the case of trichobothridial regression, the bothridial seta is absent or vestigial in one or more instars, whilst the bothridium can be reduced. The complete presence of the trichobothrium represents, in this case, the ancestral character state, the vestigial condition represents the derived character state. There are two types of trichobothridial regression in Oribatid mites: the so-called *Camisia*-type and the so-called *Hydrozetes*-type. In the *Camisia*-type, regressive evolution has started with the larva, whilst the trichobothrium can subsequently have become vestigial in protonymph, deutonymph, tritonymph and adult. A normal trichobothrium, for instance, appears in the course of postembryonic development: in the protonymph in *Archegozetes magna* (Sellnick) (Fig. 4.1 A – B); in the deutonymph in *Allonothrus schuilingi* Van der Hammen (Fig. 4.1 E – G); in the adult in *Platynothrus peltifer* (C.L. Koch); whilst it is absent in all instars in *Trhypochthoniellus setosus* Willmann (Grandjean, 1939: 304 – 306; Van der Hammen, 1955: 196, 200). Trichobothridial regression of the *Camisia*-type is known from nearly all families of the Oribatid superfamily Nothroidea (the exception is constituted by the family Hermanniidae), from the Euptyctima (superfamilies Euphthiracaroidea and Phthiracaroidea) and from two families of higher Oribatid mites (Carabodidae and Microzetidae). As is evident from Oribatid systematics (Grandjean, 1954; 1969: 139 – 152), and also from the type of manifestation, the regression cannot be reduced to a single change in character state in a common ancestor. The change must have occurred several times, independently, in a number of groups and, consequently, should be attributed to parallel evolution.

In the *Hydrozetes*-type (Fig. 4.1 C – D), the regressive evolution of the trichobothrium has started with the adult. This type is much more rare in Oribatid mites, and is known from some species of the families Hydrozetidae, Limnozetidae and Ameronothridae (Grandjean, 1949: 231; 1951: 203; Schubart, 1975: 28 – 31). Although the three Oribatid families are related, the absence of the regression in several other species of the families Hydrozetidae and Ameronothridae demonstrates that, in these families, the phenomenon must be attributed to parallel evolution.

Shifting of changes in character states, from one part of an ontogeny to another, is generally attributed to changes in systems of genotypic interactions (so-called regulatory genes are supposed to control the activity of structural

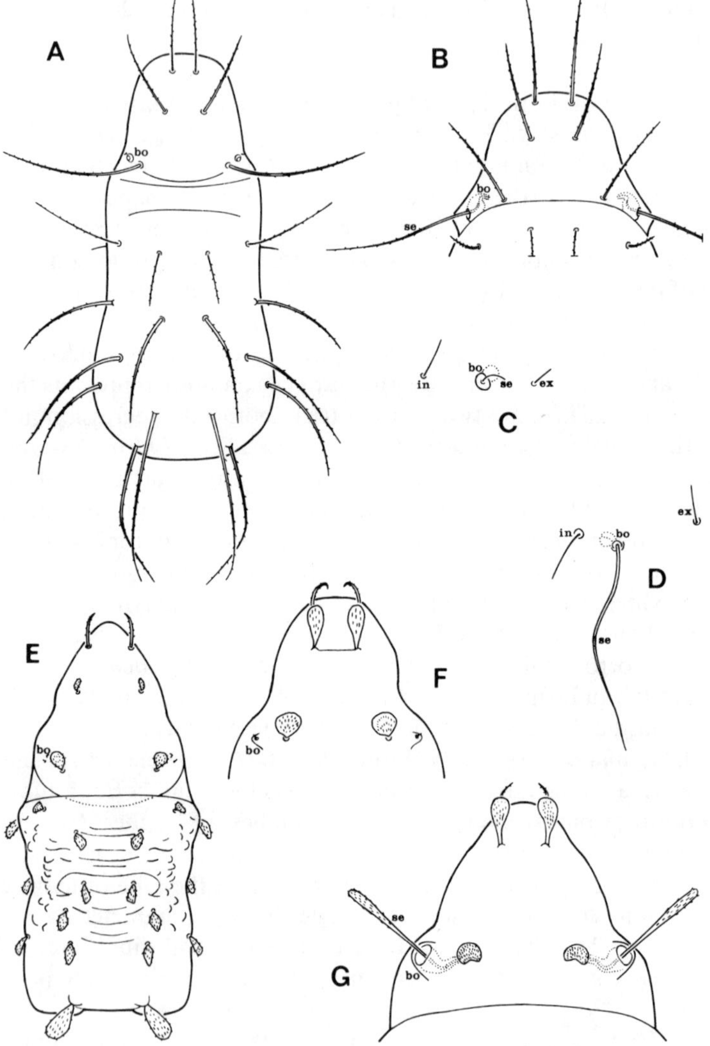

Fig. 4.1. Trichobothridial regression in Oribatid mites (A–B, E–G: ascendant type; C–D: descendant type); A–B: *Archegozetes magna* (Sellnick), dorsal view; A: larva (with vestigial trichobothrium); B: anterior part of protonymph (with normal trichobothrium); C–D: *Hydrozetes lacustris* (Michael), interlamellar seta, trichobothrium and exobothridial seta of the right side, viewed from above; C: adult (with vestigial trichobothrium); D: tritonymph (with normal trichobothrium); E–G: *Allonothrus schuilingi* Van der Hammen, dorsal view; E: larva (with vestigial trichobothrium); F: anterior part of protonymph (with vestigial trichobothrium); G: anterior part of deutonymph (with normal trichobothrium); *bo*, bothridium; *ex*, exobothridial seta; *in*, interlamellar seta; *se*, sensillus (bothridial seta); A–B: × 210; C–D: × 462, E–G: × 311.

genes). Little is known, however, about the detailed mechanisms that regulate gene activity in higher organisms (cf. Dobzhansky *et al.*, 1977: 29, 256–261, Fig. 8.11). In Arthropods, a simple change in gene regulation can manifest itself, in the course of postembryonic ontogeny, in the following three ways (cf. essay II, Fig. 2.6).

a. The replacement of an ancestral character state by a derived character state starts with one of the earlier instars and gradually manifests itself in the subsequent instars (Fig. 2.6 A, the so-called ascendant type).

b. The replacement starts with one of the later instars and gradually manifests itself in the preceding instars (Fig. 2.6 B, the so-called descendant type).

c. The replacement, when manifesting itself, always does so at the same level of postembryonic development (Fig. 2.6 C, the so-called vertical type).

The trichobothridial regression of the *Camisia*-type is ascendant, that of the *Hydrozetes*-type descendant, and because of this important difference the two regressions must be attributed to heterologous changes in systems of genotypic interactions.

Evidently, we are here concerned with homologous organs which are subject to regression under the influence of two heterologous factors. Consequently, we must distinguish two groups of parallel evolutions: (a) the trichobothridial regression of the *Camisia*-type (ascendant), which manifests itself as parallel evolution in Euphthiracaroidea, Phthiracaroidea, six families of Nothroidea, Carabodidae and Microzetidae; and (b) the trichobothridial regression of the *Hydrozetes*-type (descendant), which manifests itself as parallel evolution in Hydrozetidae, Limnozetidae and Ameronothridae. The evolutions in these two groups, when compared with each other, must be regarded as convergences.

In the study of parallel evolution, it is evidently not only of paramount importance to establish the homology of the organs in question, but also to establish whether the evolutionary changes are associated with homologies in systems of genotypic interactions.

3. Parallel evolution of homonomous elements

Similar elements in one and the same organism are generally designated as homonomous. As mentioned in the previous essays, homonomy is based on the repeated, and often developmentally varied, manifestation of the same genetic pattern. Homonomy (the condition of being homonomous) can be subdivided into general homonomy, or homonymy (e.g., the similarity, in a specimen, of its setae), and special homonomy, among which metameric homonomy (e.g., the similarity, in a specimen, of its body segments or its appendages). The evolution of the homonomous elements of an organism has often been parallel, or partly parallel, and an analysis of a few selected cases will contribute to our general understanding of the phenomenon of parallel evolution. In the present

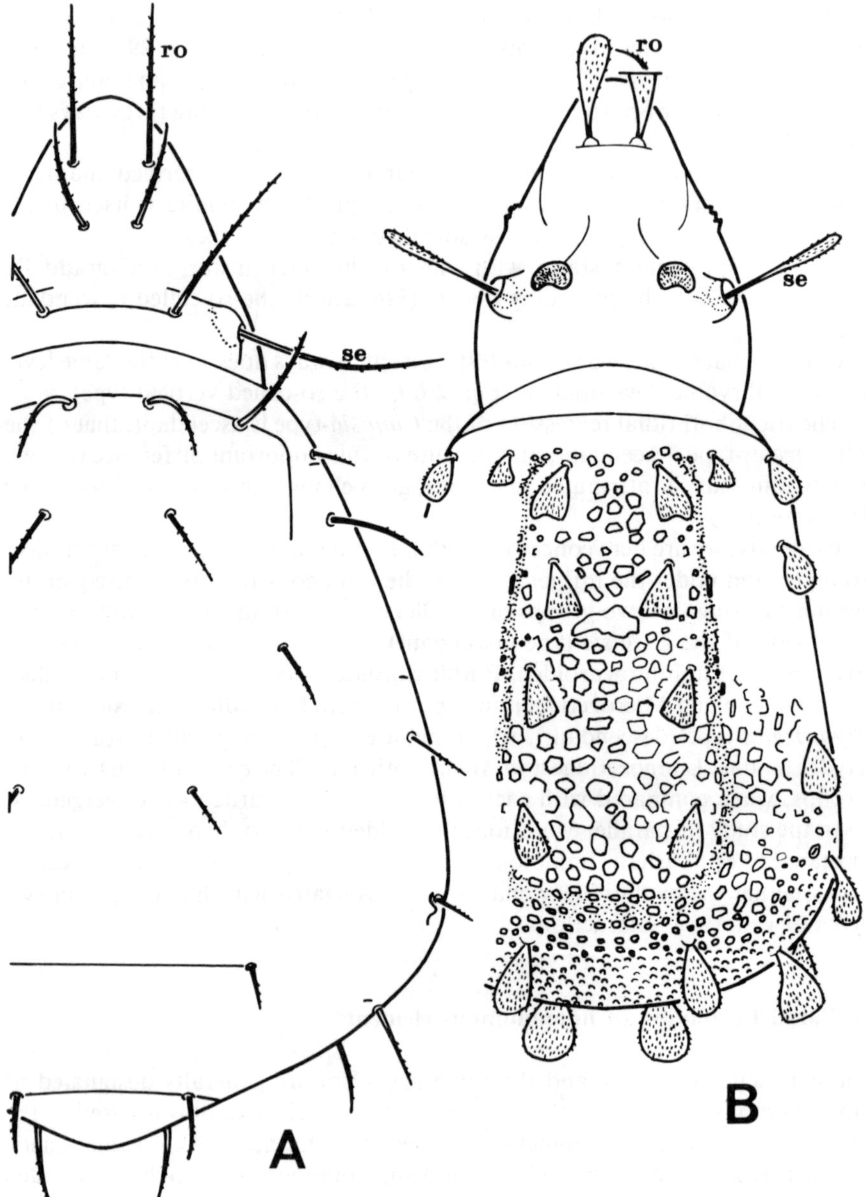

Fig. 4.2. Parallel evolution of homonomous elements: example from general homonomy (two species of the Oribatid mite family Trhypochthoniidae: one species with simple setae, and one species with fan-shaped setae); A: *Archegozetes magna* (Sellnick), adult; B: *Allonothrus schuilingi* Van der Hammen, adult; *ro*, rostral seta; *se*, sensillus (bothridial seta); A: × 136; B: × 266.

section, examples from general as well as from metameric homonomy will be examined.

In Fig. 4.2 two species of the Oribatid family Trhypochthoniidae are represented in dorsal view, viz., *Archegozetes magna* (Sellnick) and *Allonothrus schuilingi* Van der Hammen. In these species (as in most other Actinotrichid mites), the ancestral segmentation of the body has for the greater part disappeared, whilst prosoma (the anterior tagma) and opisthosoma (the posterior tagma) have been subject to divergent evolutions (as in all Chelicerata); for this reason, the setae of the body are examined here as examples from general homonomy. In most species of the family Trhypochthoniidae, the shape of the setae of prodorsum (the dorsal part of the prosoma) and notogaster (the dorsal, lateral and posterior part of the opisthosoma) is simple, as in *Archegozetes magna* (Fig. 4.2 A), or more or less clavate. In the genus *Allonothrus* (Fig. 4.2 B), however, these setae are all fan-shaped, with the exception of two pairs of prodorsal setae, viz., the rostral setae (which are simple setae) and the sensilli (which are clavate). It seems probable that the change, from simple to fan-shaped, has had a common base; in this context it is very interesting that two pairs of setae have not been included in this evolutionary program.

Fig. 4.3 A–F illustrates some aspects of parallel evolution in the case of metameric homonomy. Actinotrichid mites are ancestrally characterized by the possession of tridactyl claws (Fig. 4.3 A, C, E). In numerous cases, this tridactyl claw has been subject to regression, and can have become monodactyl by suppression of both lateral ungues, or bidactyl by suppression of either the central unguis (*oc*) or one of the lateral ungues (the anterior unguis *ol'* or the posterior unguis *ol''*). Evidently, the anterior ungues *ol'* of all legs constitute a group of metamerically homonomous elements, whilst the posterior ungues *ol''* constitute a different group of metamerically homonomous elements. When homonomous ungues of the four pairs of legs are subject to regression (as in Fig. 4.3 B, D, F, where ungues *ol''* have disappeared), the regressive evolution has been parallel; this correspondence in evolution has also been characterized as evolutionary conformity (Grandjean, 1961: 211). Absence of correspondence in evolution (or evolutionary nonconformity) is also known from the legs of Actinotrichid mites. Oribatid mites, for instance, are ancestrally characterized by the possession of lateral setae *l'* and *l''* on genu and tibia of the legs; this condition is, for instance, still found in Nothroidea (Fig. 4.3 G, J). In most Oribatid mites, however, setae *l''* have been subject to regression in legs III and IV, but not in legs I and II (Fig. 4.3 H). Consequently, the regressive evolution has been parallel for legs III and IV, but not for all legs. This evolutionary nonconformity is probably connected with the co-ordination of an internal and an external cause: several setae of the posterior face of legs III and IV, which face is in regular contact with the cuticle of the body, tend to disappear.

The evolution of the arthropod appendages (which are metamerically homonomous) constitutes a very complicated example of parallelism, in which

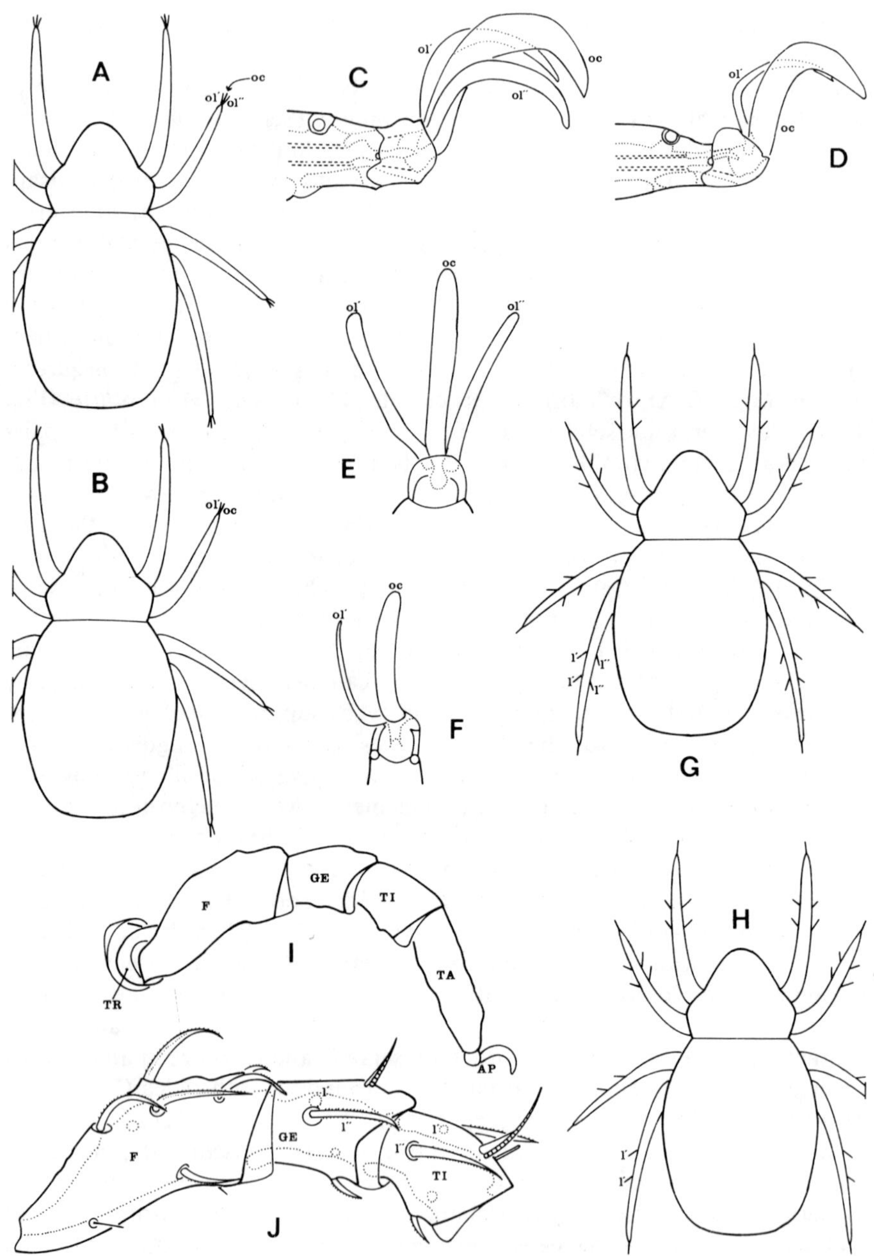

Fig. 4.3. Parallel evolution of homonomous elements (and related phenomena); examples from metameric homonomy; A, B: schematic representations of Oribatid mites in dorsal view; A: with tridactyl claws; B: with bidactyl claws (in all legs, the posterolateral unguis *ol''* has disappeared); C, D: terminal parts of right leg I, lateral view; C: *Nothrus palustris* C.L. Koch (tridactyl claw);

various stages can be distinguished, as well as a superposition of evolutionary conformous and nonconformous changes. In the following paragraph (and in Fig. 4.4) a model is prepared of the evolution of the appendages in Chelicerata.

The six pairs of chelicerate appendages (chelicera, palp and legs) are supposed to have developed from lobopodia (Fig. 4.4 A), i.e. muscular-walled, unsegmented limbs of which the rigidity was controlled by hydrostatic pressure working against the tension of muscles (Manton, 1977: 185–190, 196–198). The ancestral number of segments or podomeres of the chelicerae (three) could have been the most primitive number of segments for all the appendages (parallel evolution), and could represent the first stage in the development of segmented limbs. These three-segmented limbs or archepodia (in which the third podomere is represented by the claw-segment) could have been present in short-legged species which stood up on their legs (Fig. 4.4 B, C). Vachon (1945) demonstrated that, in chelicerate embryos, palp and legs first divide into four segments, which can be named prototrochanter, protofemur, prototibia and prototarsus. A limb with these four segments and an apotele or claw-segment (five podomeres), i.e. a so-called protopodium, could represent the first stage in the evolution of a hanging stance (Fig. 4.4 D, E). All types of chelicerate palps and legs can be derived from this primitive protopodium. It is interesting that, in this model, the evolution has been parallel for all the appendages up to the archepodium stage, and parallel for palp and legs up to the protopodium stage. The superposition of subsequent evolutionary changes is demonstrated here after the example of the Ricinulei (Fig. 4.4 F–K), a chelicerate superorder. In this group, the chelicera presents two segments (probably because of the fusion of segments 1 and 2, the so-called trochanter and the body of the chelicera). The Ricinuleid palp presents five segments (trochanter, femur, genu, tibiotarsus and apotele), but these do not all represent the prototypal podomeres because the genu arose by subdivision of the prototibia (into genu and tibia; this also occurred, by parallel evolution, in the legs), and the tibiotarsus by fusion of tibia and tarsus. In all Ricinuleid legs, a coxa arose (by parallel evolution) from the sternal and pleural regions of the body. In Ricinuleid legs III and IV a second trochanter arose by duplication of the original trochanter (this evolutionary process is analysed in section 4). In legs I–IV the prototibia has subdivided (as in the case of the palp) into genu and tibia. In all legs the tarsus has subdivided into basitarsus and telotarsus, whilst the telotarsus of the adult legs II–IV now consists of four (leg III) or five (legs II,

---

D: *Nothrus silvestris* Nicolet (bidactyl claw); E, F: terminal parts of right leg I, dorsal view; E: *Nothrus palustris* C.L. Koch; F: *Nothrus silvestris* Nicolet; G, H: schematic representations of Oribatid mites in dorsal view; G: setae *l'* and *l"* of genu and tibia present in all legs; H: setae *l"* absent in legs III and IV; I: right leg I of *Platynothrus peltifer* (C.L. Koch); J: femur, genu and tibia of right leg I of *Platynothrus peltifer* (C.L. Koch) with setae *l'* and *l"* of genu and tibia; *AP*, apotele; *F*, femur; *GE*, genu; *l'*, anterolateral seta; *l"*, posterolateral seta; *oc*, central unguis; *ol'*, anterolateral unguis; *ol"*, posterolateral unguis; *TA*, tarsus; *TI*, tibia; C–F: × 430; I: × 146; J: × 230.

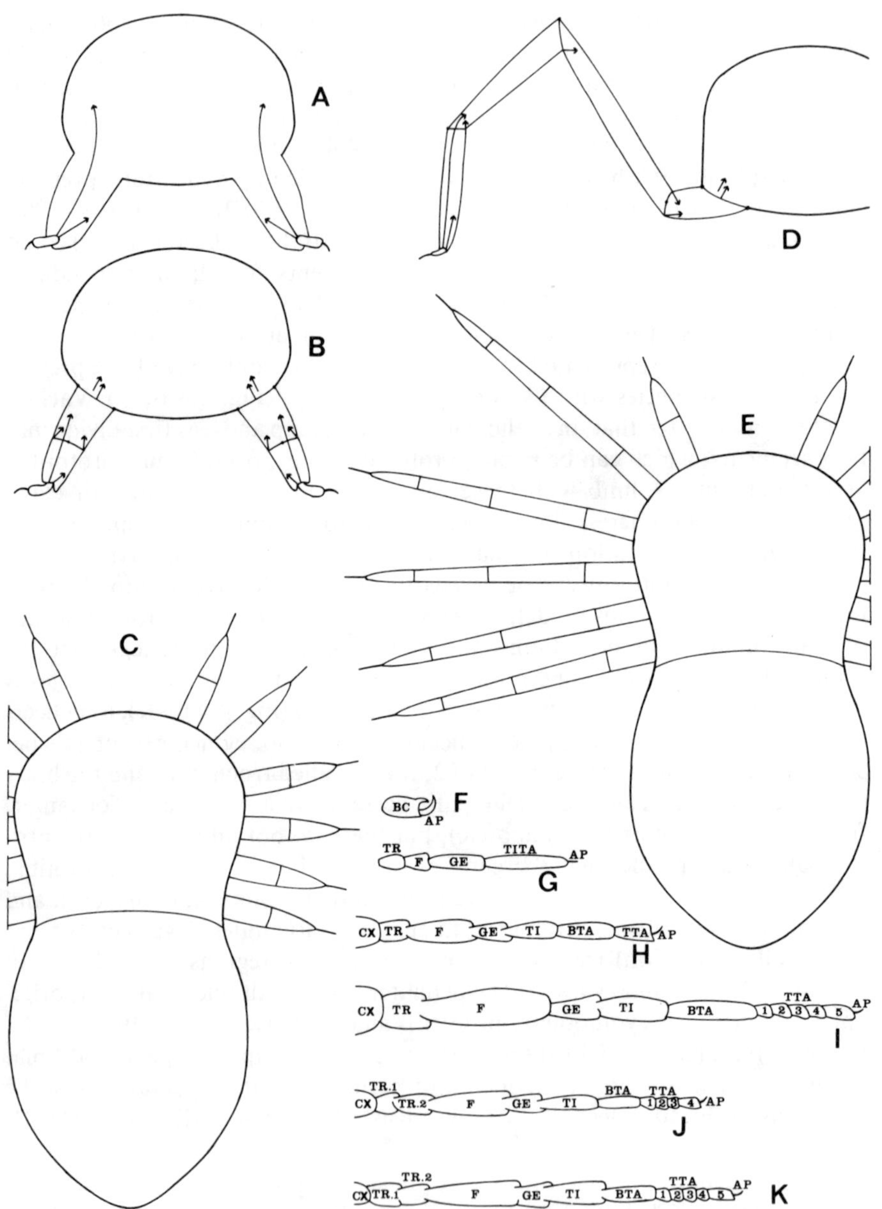

Fig. 4.4. Stages in the evolution of arthropod appendages; A, B: single body-segment, with pair of limbs, of hypothetical Arthropods in transverse section; A: with lobopodia; B: with archepodia; C: dorsal view of hypothetical Arthropod with archepodia; D: single body-segment, with left limb, in transverse section, of hypothetical Arthropod with protopodia; E: dorsal view of hypothetical Arthropod with three-segmented limb 1 and five-segmented protopodia (limbs 2−6); F−K: ap-

IV) adesmatic segments (which probably arose by subdivision and subsequent multiplication). This example clearly demonstrates the complicateness of the evolutionary pattern underlying the morphology of the Ricinuleid appendages, which pattern consists of a superimposition of parallel and divergent evolutions.

As a final example of the parallel evolution of homonomous elements, mention can be made of the postembryonic development of an Arthropod with, ancestrally, a series of similar immature instars (the result of parallel evolution) constituting a single phase. In the beginning of arthropod evolution, parallel evolution of instars must have been a general phenomenon; it was not until a subsequent divergent evolution (superimposed on the original parallelism) that radical metamorphosis arose (cf. essay III).

4. Parallelisms in the evolution of chelicerate legs

The last two sections of the present essay are concerned with parallelisms as fundamental problems in the higher classification of Chelicerata. It is supposed now (Van der Hammen, 1985a) that the main chelicerate groups were already separated by important discontinuities in the Cambrian, and that the fundamental differences are connected with specializations and adaptations in the field of locomotion and ingestion. A survey of the present view of chelicerate phylogenetic relationships is given in Fig. 4.5; this diagram still contains a number of uncertainties (the dotted lines), which are caused by the great difficulties connected with the reconstruction of Cambrian and Precambrian branching points. The present section is devoted to parallelisms in the evolution of chelicerate legs, and links up with part of section 3 (in which the general evolution of chelicerate limbs, from lobopodia to archepodia and protopodia is discussed). Starting from the protopodium, the evolution of chelicerate legs presents several parallelisms as well as divergencies; an analysis of these phenomena necessitates the following survey of the segmentation and articulation of the chelicerate legs (Van der Hammen, 1985, 1985a, 1986a).

Arthropod legs present several types of joints: adesmatic joints (i.e. joints without tendons inserted on the base of the distal segment) and various eudesmatic joints (i.e. joints with one or more tendons inserted on the base of the distal segment). Generally, two basic types of eudesmatic joints can be distinguished, and two derived types. The two basic types of eudesmatic joints are hinge joint (with one articulation point and one set of tendons and muscles) and pivot joint (with two articulation points and two sets of antagonistic ten-

---

pendages of adult Ricinulei in dorsal view; F: chelicera; G: palp; H: leg I; I: leg II; J: leg III; K: leg IV; *AP*: apotele; *BC*: body of chelicera; *BTA*: basitarsus; *CX*: coxa; *F*, femur; *GE*: genu; *TI*: tibia; *TITA*: tibiotarsus; *TR*: trochanter; *TR.1*: trochanter 1; *TR.2*, trochanter 2; *TTA*: telotarsus; *TTA.1–5*: telotarsus 1–5.

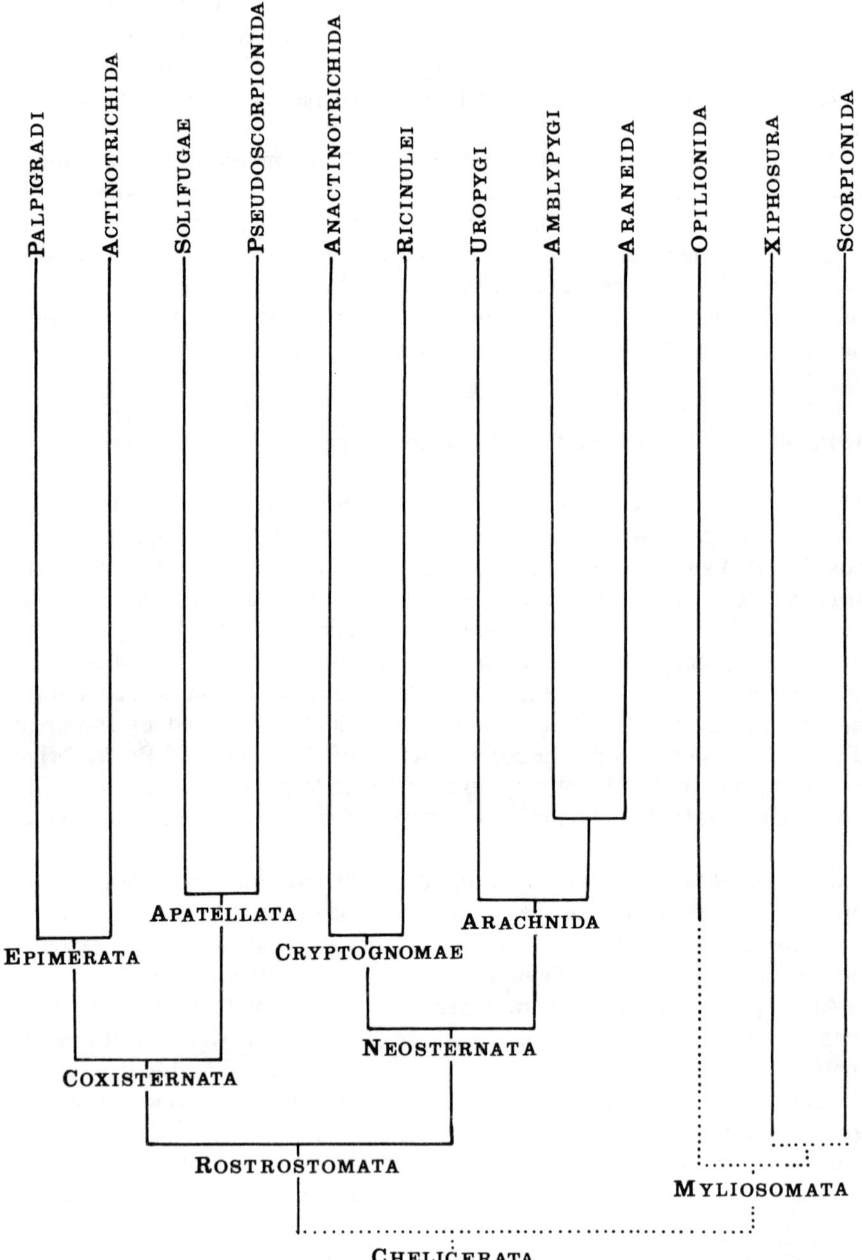

Fig. 4.5. Diagrammatic representation of the present view of chelicerate phylogenetic relationships. The names of three taxa (the classes Coxisternata and Neosternata, and the superclass Rostrostomata) are new; they will be defined in An Introduction to Comparative Arachnology, my book on Chelicerata, which will be published before long.

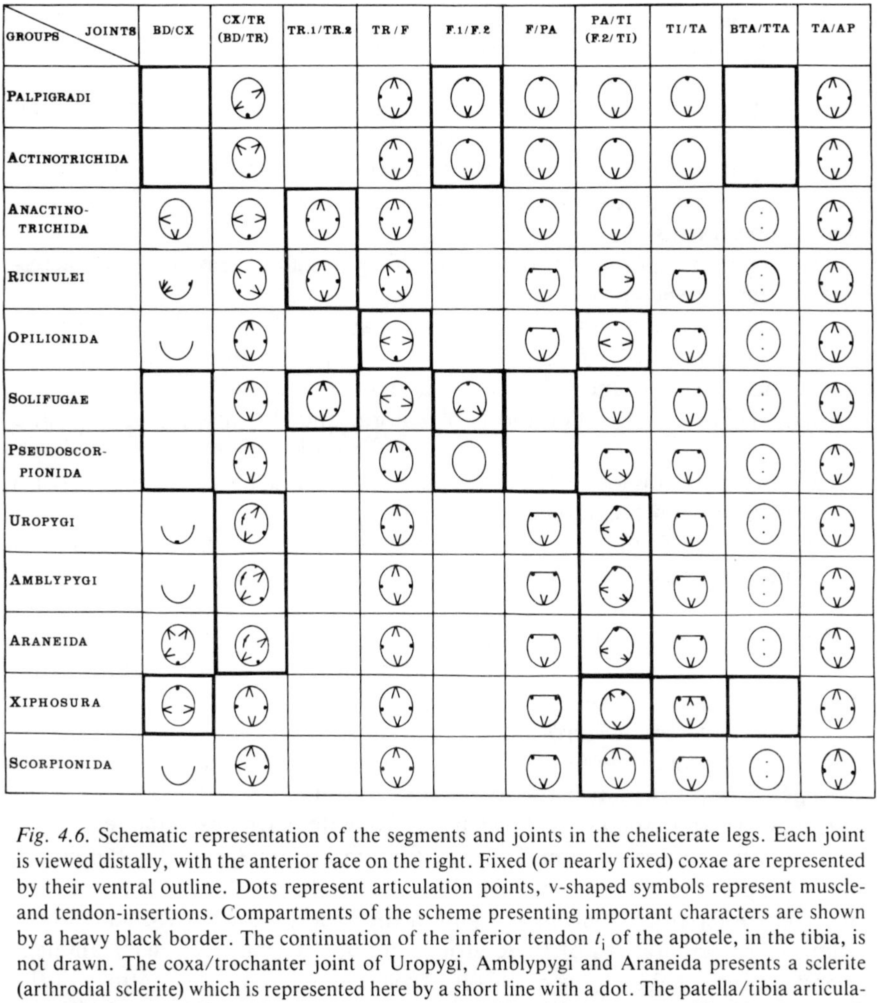

GROUPS \ JOINTS	BD/CX	CX/TR (BD/TR)	TR.1/TR.2	TR/F	F.1/F.2	F/PA	PA/TI (F.2/TI)	TI/TA	BTA/TTA	TA/AP
PALPIGRADI										
ACTINOTRICHIDA										
ANACTINO-TRICHIDA										
RICINULEI										
OPILIONIDA										
SOLIFUGAE										
PSEUDOSCOR-PIONIDA										
UROPYGI										
AMBLYPYGI										
ARANEIDA										
XIPHOSURA										
SCORPIONIDA										

Fig. 4.6. Schematic representation of the segments and joints in the chelicerate legs. Each joint is viewed distally, with the anterior face on the right. Fixed (or nearly fixed) coxae are represented by their ventral outline. Dots represent articulation points, v-shaped symbols represent muscle- and tendon-insertions. Compartments of the scheme presenting important characters are shown by a heavy black border. The continuation of the inferior tendon t_i of the apotele, in the tibia, is not drawn. The coxa/trochanter joint of Uropygi, Amblypygi and Araneida presents a sclerite (arthrodial sclerite) which is represented here by a short line with a dot. The patella/tibia articulation of Uropygi and Araneida can be much more complicated than represented in this scheme (additional condyle, additional traversing muscle); the patella/tibia articulation of Amblypygi permits very little mobility. The scheme is based on right leg IV, and represents the fundamental leg-type of each group. Abbreviations: *AP*: apotele; *BD*: body; *BTA*: basitarsus; *CX*: coxa; *F*: femur; *F.1*: femur 1; *F.2*: femur 2; *PA*: patella (or genu); *TA*: tarsus; *TI*: tibia; *TR*: trochanter; *TR.1*: trochanter 1; *TR.2*: trochanter 2; *TTA*: telotarsus.

dons and muscles). When, in a pivot joint, one of the articulation points disappears, a rocking joint arises, which permits leg rocking. In some cases, extensor muscles can arise in a hinge joint; this is particularly found in the cases of appendages which are used as palp. Pivot joints are either associated with levator-depressor, or with promotor-remotor movements. Hinge joints are associated with flexure of the legs. The distribution of these types of joints in the legs of various groups of Chelicerata presents a degree of diversity not found in other Arthropod groups. In Fig. 4.6 a general survey is given of this distribution in the fundamental leg types of each of the main groups of Chelicerata. The figure demonstrates that, in extant Chelicerata, the following eight fundamental types of legs are present (subsequent changes in these types are not considered).

The first type, found in Palpigradi and Actinotrichid mites, is characterized by the absence of a coxa, the presence of two femora, and the absence of a separate basitarsus.

The second type, found in Anactinotrichid mites and Ricinulei, is characterized by the presence of two trochanters in legs III and IV.

The third type, found in Opilionida, is characterized by the presence of rocking joints between trochanter and femur, and between patella and tibia.

The fourth type, found in Solifugae, is characterized by the presence of two trochanters (in legs III–IV) and two femora, and the absence of coxa and patella.

The fifth type, found in Pseudoscorpionida, is characterized by the presence of two femora and the absence of coxa and patella.

The sixth type, found in Uropygi, Amblypygi and Araneida, is characterized by the presence of specialized rocking joints between coxa and trochanter, and between patella and tibia.

The seventh type, found in Xiphosura, is characterized by the presence of an articulation between body and coxa, which (in contradistinction to all other extant Chelicerata) is associated with the promotor-remotor swing of the leg; in this type, there is no separate basitarsus.

The eighth type, found in Scorpionida, is particularly characterized by the presence of a pivot joint between patella and tibia.

Each of these fundamental types must originally have been adapted to one particular type of locomotion. In the case of the first type, found in Palpigradi and Actinotrichid mites, ancestral locomotion could have been associated with life in the interstitial environment and in the deep fissures of the soil, where primitive representatives of these groups are now found (Van der Hammen, 1985); a functional analysis of this leg type is, however, still wanting. In the case of the second type, with two trochanters in legs III and IV, ancestral locomotion could have been walking as well as swimming; legs III and IV could have been used, as in the fossil *Baltoeurypterus* (cf. Selden, 1981), for swimming.

A comparison of the eight fundamental leg-types (Fig. 4.6) with the protopodium (the hypothetical leg-type common to all chelicerate ancestors), represented in Fig. 4.4 D, demonstrates which changes must have taken place during the evolution of the main groups. The most important of these changes are the origination of a coxa, a second trochanter, a second femur, a patella or genu, and a separate basitarsus. As mentioned in section 3, a coxa is supposed to have arisen from sternal and pleural regions of the body, at the base of the leg. In Anactinotrichid mites, trochanter 2 disappears as a result of sup-

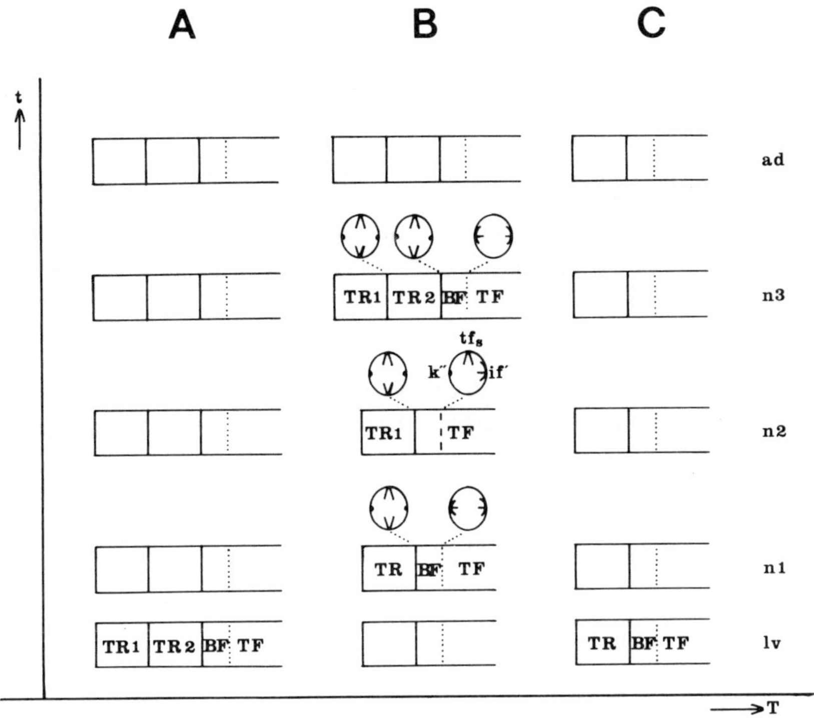

Fig. 4.7. Ontophylogenetic diagram (not drawn to scale) of the regressive evolution of trochanter 2 in Anactinotrichida. Trochanter 2 disappears by ascendant suppression. In Opilioacarid deutonymphs, the regressive trochanter 2 manifests itself partly (the 'basifemoral ring' presents some characters of the trochanter 2 / femur articulation, viz., the superior tendon and the posterior condyle; and some characters of a basifemoral ring, viz., an anterior lyrifissure and the absence of an inferior tendon). The suppression is attributed to the regulatory mechanisms of the genome. A: ancestral Anactinotrichid stage (two trochanters in all instars are still found in other Cryptognomae, viz., Ricinulei, in which group, however, a basifemoral ring is not present); B: Opilioacarid stage; C: Holothyrid, Gamasid and Ixodid stage. Abbreviations: *ad*: adult; *BF*: basifemur; *if'*: anterior lyrifissure of basifemur; *k"*: posterior condyle of femur; *lv*: larva; *n1*: protonymph; *n2*: deutonymph; *n3*: tritonymph; *t*: ontogenetic time; *T*: phylogenetic time; *TF*: telofemur; tf_s: superior tendon of femur; *TR*: trochanter; *TR1*: trochanter 1; *TR2*: trochanter 2.

pression (Fig. 4.7) (Van der Hammen, 1985), and it must originally have evolved by duplication (repetition of information). In Actinotrichid mites, femur 2 disappears by fusion with femur 1 (Van der Hammen, 1982: 38), and it must originally have evolved by subdivision of the protofemur. In the same way, the patella (or genu) arose by subdivision of the prototibia, and the basitarsus by subdivision of the prototarsus.

When we compare the survey of the segments and joints in the chelicerate legs (Fig. 4.6) with the diagram of the phylogenetic relationships (Fig. 4.5), it becomes evident which cases are attributable to parallelisms (the phylogenetic relationships are discussed by Van der Hammen, 1977, 1986a, 1986c). A coxa is present in the legs of all Chelicerata, except Coxisternata (Epimerata and Apatellata); from the phylogenetic diagram, it can be deduced that the presence of this podomere must have developed in Neosternata and Myliosomata by parallel evolution. From the diagram it is also evident that the second trochanter of legs III and IV in Cryptognomae and Solifugae must be regarded as the result of parallelism. The evolution of a genu or patella in most groups of Chelicerata (except Apatellata) could be attributable to two different parallelisms (because of important differences in the articulation) A separate basitarsus is present in all Chelicerata except Coxisternata and Xiphosura; it must have evolved by parallel evolution in Apatellata, Neosternata, Opilionida and Scorpionida. In Palpigradi and some Actinotrichida (Actinedida), the tarsus is subdivided into a number of adesmatic tarsal podomeres (there is no subdivision into basi- and telotarsus); this evolution can also be regarded as parallel.

5. Parallelisms in the evolution of the gnathosoma

A so-called gnathosoma (a cone-shaped anterior division of the body, a pseudotagma, which includes the mouthparts, and which has secondarily become mobile) is present in three superorders of Chelicerata: Actinotrichida, Anactinotrichida and Ricinulei. The last-mentioned two groups are related (they constitute the Cryptognomae, a subclass). Actinotrichida, however, are related to Palpigradi (the two groups constitute the Epimerata, another subclass). As will be evident from Fig. 4.5, these subclasses are related to different other subclasses (Epimerata to Apatellata, Cryptognomae to Arachnida), and because of this it will be interesting to analyse the parallelisms and convergences in the evolution of the gnathosoma. It may be remarked here that, particularly because of the presence of a gnathosoma, Actinotrichida and Anactinotrichida have for a long time been regarded as constituting the monophyletic group of the Acari or mites. Detailed comparative studies (Van der Hammen, 1972, 1977, 1979, 1982, 1986c), however, have gradually revealed fundamental morphological differences, and different relationships.

$\longrightarrow$

Fig. 4.8. Schematic representation of the mouthparts of acaromorph Chelicerata; A: gnathosoma of Actinotrichida, lateral view; B: mouthparts of Palpigradi, lateral view; C: gnathosoma of Opilioacarida (Anactinotrichida), ventral view; D: gnathosoma of Anactinotrichida (lateral view). *Ac.P:*, acetabulum of palp; *Ac.I*: acetabulum of leg I; *b*: mouth; *CH*: chelicera; *CO*: corniculus; *cpc*: podocephalic canal; *CX.I*: coxa of leg I; *CX.II*: coxa of leg II; *e*: supracoxal seta of palp; *eI*: supracoxal seta of leg I; *H*: mentum; *Ji*: inferior commissure of mouth; *Js'*: right superior commissure of mouth; *LI*: labium; *LL*: lateral lip; *LS*: labrum; *ogc*: orifice of coxal gland; *PH*: pharynx; *RU*: rutellum; *SA*: sternapophysis; *sg*: subcapitular gutter; *s.t*: sternal taenidium; *TR.P*: trochanter of palp; *TR.I*: trochanter of leg I.

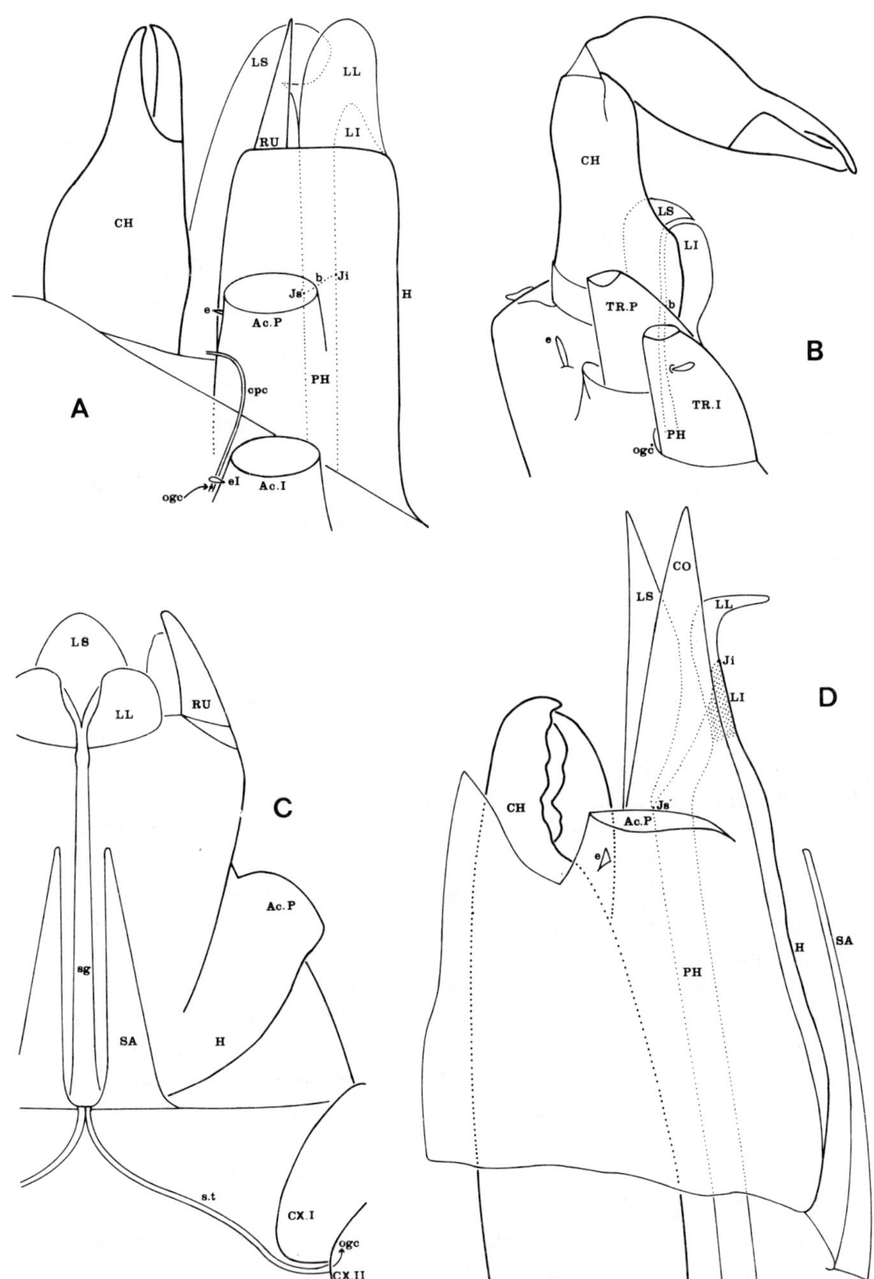
LS
LL
RU
LI
CH
Ji
b
Js
e
Ac.P
H
cpc
PH
A
el
ogc
Ac.I
CH
LS
LI
TR.P
b
e
TR.I
PH
ogc
B
LS
LL
RU
Ac.P
sg
SA
H
s.t
CX.I
ogc
CX.II
C
CO
LS
LL
Ji
LI
CH
Js
Ac.P
e
PH
H
SA
D

98

In the present section, a comparative study is made of the mouthparts of both groups of Epimerata and both groups of Cryptognomae. The structures are schematically represented in Fig. 4.8. They are first studied separately; a subsequent comparative analysis will put us in a position to draw some conclusions concerning parallelism and convergence. Some passing remarks are made on the mouthparts of Apatellata and Arachnida.

The mouthparts of Palpigradi (Fig. 4.8 B) (see also: Van der Hammen, 1969; 1982: 13, Fig. 4) include labrum, labium, the crescent-shaped mouth (with two commissures), the pharynx and the chelicerae; the Palpigradid palp has an ambulatory function, and its coxisternal region or epimeron has fused with the epimeron of leg I. A fusion of these epimera is also known from the Oribatid mite family Epilohmanniidae (Van der Hammen, 1982: 34, Fig. 15). A detailed comparative study of the Epilohmanniid gnathosoma (it has lost its special gnatho-idiosomatic articulation; part of the role of this articulation is apparently taken over by the so-called labiogenal articulation) and the Palpigradid mouthparts has recently been made by me (Van der Hammen, 1986b). It appears that the fusion of the two epimeral regions, in both groups, represents different adaptations (in Palpigradi, it could pertain to the functional association of palp and leg I). The orifice of the Palpigradid coxal gland is above the base of leg I; there could be a narrow, open canal, extending from this orifice to the mouth, although it has not yet been discerned with certainty.

The gnathosoma of Actinotrichida, Anactinotrichida and Ricinulei is (with few exceptions of secondary origin) movably attached to the idiosoma, and consists of cheliceral frame and infracapitulum. The cheliceral frame constitutes the body wall between the rostral region of the prosoma and the infracapitulum; the chelicerae are attached to it in a movable way. It comprises a superior, a narrow median, and an inferior region. The infracapitulum constitutes the inferior part of the gnathosoma, bearing lips and palpi, and containing mouth and pharynx. Generally, the following regions can be distinguished: mentum (the coxisternal region, or epimeron, of the palp), cervix (the dorsal region, which constitutes the base of the labrum), and the large lateral ridges (with the acetabulum of the palp). There can be two apodemes: a capitular apodeme which is the internal continuation of the cervix, and a subcapitular apodeme which is the internal continuation of the mentum; muscles can also be inserted on the dorsal and lateral parts of the gnathosomatic base. The orifice of the coxal gland can be associated with the gnathosoma in various ways.

The Actinotrichid gnathosoma (see Fig. 4.8 A; and Van der Hammen, 1968; 1982: 34–37, Figs. 16–18) is, ancestrally, particularly characterized by its association with a podocephalic canal which (originally) extends from the orifice of the coxal gland (above leg I) to the cervix (the dorsal region of the infracapitulum). Ancestrally, there are four lips (labrum, lateral lips and labium) and two apodemes (capitular and subcapitular apodeme); labium and subcapitular apodeme have disappeared in most higher Actinotrichid mites. Articulation of the gnathosoma (the articulation point is constituted by a socket, or glenoid cavity, in the posterior part of the large lateral ridges, and an idiosomatic condyle) is operated by muscles inserted on the capitular apodeme (and on the subcapitular apodeme, in case this is present).

The Anactinotrichid gnathosoma (see Fig. 4.8 C, D; and Van der Hammen, 1961: 185−189, Figs. 2−4, 5 C; 1964: 12−25, Figs. 5−12; 1965: 258−269, Figs. 2 B, C, 3−7; 1966: 23−32, Figs. 7−12; 1983: 218−224, Figs. 7−11; 1986c: 220−223, Fig. 2) is, ancestrally, particularly characterized by the presence of a subcapitular gutter (with ventral position), associated with the orifices of the coxal glands (with ventral position, posterior to coxae I) and with the sternapophyses. Generally, there are three lips, viz., labrum and lateral lips (the labium is probably incorporated in the mentum; the inferior commissure of the mouth has an advanced position), and a subcapitular apodeme (there is no capitular apodeme); intrinsic muscles of the gnathosoma can be attached to the internal continuation of the cervix. The superior part of the cheliceral frame is extended in anterior direction. Movement of the gnathosoma is operated by muscles inserted on the subcapitular apodeme and on the lateral and superior parts of the gnathosomatic base (there is no gnatho-idiosomatic articulation point).

The Ricinuleid gnathosoma (Van der Hammen, 1979: 9−10, Figs. 5−7) presents a subcapitular gutter (with ventral position) associated with the orifices of the coxal glands (with ventral position, paraxial to coxae I) and with the sternapophyses. It is, moreover, characterized by the presence of a subcapitular apodeme, the presence of muscles inserted in the lateral part of the gnathosomatic base, the absence of lateral lips, and the complete incorporation of the labium in the mentum (which is extended in anterior direction). In its essential characters, the Ricinuleid gnathosoma is closely related to the Anactinotrichid gnathosoma, although it is more primitive (absence of lateral lips).

It is evident that the morphology of the gnathosoma in Actinotrichida, Anactinotrichida and Ricinulei presents indeed a number of striking similarities, particularly with reference to the infracapitulum (cervix with labrum, and attachment of cheliceral frame; mentum; lateral ridges with acetabulum of palp), which, in the case of Actinotrichida and Anactinotrichida, previously led to the conception of the mites as a monophyletic group (the occurrence of a gnathosoma in Ricinulei was not recognized until recently; see Van der Hammen, 1979). As I have demonstrated before (Van der Hammen, 1971: 472; 1972: 274), the great differences in association with the orifices of the coxal glands (in Actinotrichida, the orifice is, ancestrally, on either side, dorsal to trochanter I, and connected by a podocephalic canal with the dorsal cervix; in Cryptognomae, i.e., Anactinotrichida and Ricinulei, the orifice is, ancestrally, on either side, posteroventral to coxa I, and associated with a ventral subcapitular gutter and with sternapophyses), indicate that the evolution of the gnathosoma, in these two groups, must have been separate from the beginning. The different origin of the gnathosoma in the two groups is also evident from the following data: (a) a capitular apodeme, functioning as the place of insertion of extrinsic muscles, and present in Actinotrichida, is also found in Apatellata; (b) an interior prolongation of the cervix, functioning as the place of attachment of intrinsic muscles, and present in Cryptognomae, is also found in Arachnida.

Because the main parts of the gnathosoma, in Actinotrichida and Cryptognomae, evidently arose from homologous structural elements, and because of great similarities in general architecture, the process of gnathosomatization in the two groups could, notwithstanding the fundamental differences, be characterized as an example of partial parallelism.

6. Conclusions

The special examples, analysed in the previous sections, enable us to draw a few conclusions of a more general character. From the example of the trichobothridial regression in Oribatid mites, it appears that similar changes in the character state of homologous elements can manifest themselves, in different ways, under the influence of different regulatory mechanisms. Only those changes which are associated with homologous systems of genotypic interactions can be characterized as either shared derived attributes or parallelisms, whilst those associated with heterologous systems should be regarded as convergences. The analysis of the parallel evolutions of homonomous elements demonstrates the possible part of collective coding (and the possibility of exclusion of special elements), and the occurrence of complicated superpositions of parallel and divergent evolutions. In the analysis of the evolution of the chelicerate legs, parallelisms in the evolution of the fundamental leg-types are attributed to three different evolutionary processes: the separation of the coxa from the prosoma, duplication or multiplication of segments (repetition of information), and subdivision. In the comparative study of the gnathosoma, it is demonstrated that, in the case of complicated structures, parallelism can be partial, and the structures in question composed of homologous and heterologous elements. It may, generally, be concluded that parallelisms are founded on homologous genotypic potentialities, whilst the manifestation is attributable to homologous changes in systems of genotypic interactions.[1]

References

Camin, J.H. & R. Sokal, 1965. A method for deducing branching sequences in phylogeny. – Evolution 19: 311–326, Figs. 1–4, Tab. 1–6.

---

[1] This is in accordance with the results obtained by Gould (1984) in the analysis of dwarfing and gigantism in Gastropod shells of the genus *Cerion*. The phenomena, mentioned by Gould as convergences, are here regarded as parallelisms.

Dobzhansky, Th., F.J. Ayala, G.L. Stebbins & J.W. Valentine, 1977. Evolution. – San Francisco (W.H. Freeman and Company): xiv + 572 p.

Gosliner, T.M. & M.T. Ghiselin, 1984. Parallel evolution in opisthobranch Gastropods and its implications for phylogenetic methodology. – Syst. Zool. 33: 255–274, Figs. 1–5.

Gould, S.J., 1984. Morphological channeling by structural constraint: convergence in styles of dwarfing and gigantism in Cerion, with a description of two new fossil species and a report on the discovery of the largest Cerion. – Paleobiology 10: 172–194, Figs. 1–14, Tab. 1–8.

Grandjean, F., 1939. Observations sur les Oribates (12e série). – Bull. Mus. Nat. Hist. Natur. (2) 11: 300–307, Figs. A–G.

Grandjean, F., 1949. Sur le genre Hydrozetes Berl. (Acariens). – Bull. Mus. Nat. Hist. Natur. (2) 21: 224–231.

Grandjean, F., 1951. Comparaison du genre Limnozetes au genre Hydrozetes (Oribates). – Bull. Mus. Nat. Hist. Natur. (2) 23: 200–207, Figs. A–G.

Grandjean, F., 1954. Essai de classification des Oribates (Acariens). – Bull. Soc. Zool. France 78: 421–446.

Grandjean, F., 1961. Nouvelles observations sur les Oribates (1re série). – Acarologia 3: 206–231.

Grandjean, F., 1969. Considérations sur le classement des Oribates. Leur division en 6 groupes majeurs. – Acarologia 11: 127–153.

Hammen, L. van der, 1955. Notes on the Oribatei (Acari) of Dutch New Guinea III. The development of Archegozetes magna (Sellnick) and Allonothrus schuilingi (Van der Hammen). – Proc. Kon. Ned. Akad. Wet. (C) 58: 195–205, Figs. 1–8.

Hammen, L. van der, 1961. Description of Holothyrus grandjeani nov. spec., and notes on the classification of the mites. – Nova Guinea, Zool. 9: 173–194, Figs. 1–9, Pl. 6.

Hammen, L. van der, 1964. The morphology of Glyptholaspis confusa (Foà, 1900) (Acarida, Gamasina). – Zool. Verh. Leiden 71: 1–56, Figs. 1–21.

Hammen, L. van der, 1965. Further notes on the Holothyrina (Acarida) I. Supplementary description of Holothyrus coccinella Gervais. – Zool. Meded. Leiden 40: 253–276, Figs. 1–9.

Hammen, L. van der, 1966. Studies on Opilioacarida (Arachnida) I. Description of Opilioacarus texanus (Chamberlin & Mulaik) and revised classification of the genera. – Zool. Verh. Leiden 86: 1–80, Figs. 1–21.

Hammen, L. van der, 1968. The gnathosoma of Hermannia convexa (C.L. Koch) (Acarida: Oribatina) and comparative remarks on its morphology in other mites. – Zool. Verh. Leiden 93: 1–45, Figs. 1–12.

Hammen, L. van der, 1969. Notes on the mouthparts of Eukoenenia mirabilis (Grassi) (Arachnidea: Palpigradi). – Zool. Meded. Leiden 44: 41–45, Fig. 1.

Hammen, L. van der, 1971. La phylogénèse des Oplioacarides, et leurs affinités avec les autres Acariens. – Acarologia 12: 465–473, Figs. 1–5.

Hammen, L. van der, 1972. A revised classification of the mites (Arachnidea, Acarida) with diagnoses, a key, and notes on phylogeny. – Zool. Meded. Leiden 47: 273–292, Fig. 1.

Hammen, L. van der, 1977. A new classification of Chelicerata. – Zool. Meded. Leiden 51: 307–319, Fig. 1, Tab. 1–3.

Hammen, L. van der, 1979. Comparative studies in Chelicerata I. The Cryptognomae (Ricinulei, Architarbi and Anactinotrichida). – Zool. Verh. Leiden 174: 1–62, Figs. 1–31.

Hammen, L. van der, 1981. Numerical changes and evolution in Actinotrichid mites (Chelicerata). – Zool. Verh. Leiden 182: 1–47, Figs. 1–13. [Cf. essay II].

Hammen, L. van der, 1982. Comparative studies in Chelicerata II. Epimerata (Palpigradi and Actinotrichida). – Zool. Verh. Leiden 196: 1–70, Figs. 1–31.

Hammen, L. van der, 1983. Notes on the comparative morphology of ticks (Anactinotrichida: Ixodida). – Zool. Meded. Leiden 57: 209–242, Figs. 1–23.

Hammen, L. van der, 1985. Functional morphology and affinities of extant Chelicerata in evolutionary perspective. – Trans. Roy. Soc. Edinburgh, Earth Sciences 76: 137–146, Figs. 1–8.

Hammen, L. van der, 1985a. Comparative studies in Chelicerata III. Opilionida. – Zool. Verh. Leiden 220: 1–60, Figs. 1–34.

Hammen, L. van der, 1986. On some aspects of parallel evolution in Chelicerata. – Acta Bio-theoretica 35: 15–37, Figs. 1–9.

Hammen, L. van der, 1986a. Comparative studies in Chelicerata IV. Apatellata, Arachnida, Scorpionida, Xiphosura. – Zool. Verh. Leiden 226: 1–52, Figs. 1–23.

Hammen, L. van der, 1986b. The morphology of Epilohmannia zwarti spec. nov., an Oribatid mite from New Guinea. – Zool. Meded. Leiden 60: 71–85, Figs. 1–5.

Hammen, L. van der, 1986c. Acarological and arachnological notes. – Zool. Meded. Leiden 60: 217–230, Figs. 1–3.

Manton, S.M., 1977. The Arthropoda. Habits, functional morphology, and evolution. – Oxford (Clarendon Press): xxii + 527 p., 186 figs., 8 pls.

Mayr, E. 1969. Principles of systematic zoology. – New York (McGraw-Hill Book Company): xiv + 428 p., figs., tab.

Remane, A., 1956. Die Grundlagen des natürlichen Systems, der vergleichenden Anatomie und der Phylogenetik. Theoretische Morphologie und Systematik I. – Leipzig (second edition) (Akademische Verlagsgesellschaft Geest & Portig K.-G.): vi + 364 p., 82 figs. [I have used the reprint edition, Koenigstein-Taunus (Otto Koeltz), 1971].

Schubart, H., 1975. Morphologische Grundlagen für die Klärung der Verwandtschaftsbeziehungen innerhalb der Milbenfamilie Ameronothridae (Acari, Oribatei). – Zoologica, Stuttgart 123: 23–91, Figs. 1–43, Pls. 1–3.

Selden, P.A., 1981. Functional morphology of the prosoma of Baltoeurypterus tetragonophthalmus (Fischer) (Chelicerata: Eurypterida). – Trans. Roy. Soc. Edinburgh, Earth Sciences 72: 9–48, Figs. 1–32.

Simpson, G.G., 1961. Principles of animal taxonomy. – New York (Columbia University Press): xiv + 247 p., 30 figs.

Sneath, P.H.A. & R.R. Sokal, 1973. Numerical taxonomy. The principles and practice of numerical classification. – San Francisco (W.H. Freeman and Company): xvi + 573 p., figs., tab.

Vachon, M., 1945. L'appendice arachnidien et son évolution. Note préliminaire. – Bull. Soc. Zool. France 69: 172–177.

V. TYPE-CONCEPT, HIGHER CLASSIFICATION AND EVOLUTION

1. Introduction

An earlier version of the present essay was published several years ago in Acta Biotheoretica (Van der Hammen, 1981). It is now revised and adapted, to be included in the present book, and leads up to essay VI which is devoted to structuralism.

The type-concept is frequently used in the life-sciences and humanities, both in the sense of model, pattern, standard, norm, and in the sense of ordering category with imprecise, uncertain boundary. Curiously, it stands in bad repute in taxonomy and morphology, owing to some strange remarks on the subject made in literature. Characterizing an investigator as a typologist here is nearly the same as using a term of abuse. It has erroneously been supposed that typology stems from Plato, and that a type is identical with a Platonic Idea (Mayr, Linsley & Usinger, 1953: 15; Simpson, 1967: 46; Mayr, 1969: 24, 66). Some authors have indeed wrongly claimed a Platonic origin for their idealistic morphology (Troll, 1928; 1941; Meyer-Abich, 1934; 1963). As will be demonstrated below, the type-concept as first used in biology in the eighteenth century, stems from Leibniz's concept of substantial form, and could now be understood in the sense of genotypic pattern. As mentioned already in the introduction, one of the essential characters of the eighteenth- and nineteenth-century type-concept was constituted by its being composed of homonomous parts (homotypes); according to recent discoveries in morphogenetics and molecular cytogenetics, these can now be regarded as developmental (and evolutionary) variations of a single basic genetic pattern.

The type-concept is developed here as a model of evolution and a standard of higher classification. In the course of my studies of the classification of Chelicerata (Van der Hammen, 1977, 1979, 1982, 1985, 1986a, 1986b), and on evolutionary potentialities, evolutionary mechanisms, and parallelisms in manifestation (Van der Hammen, 1981a, 1986), it became increasingly obvious that a higher classification should be based on a hierarchic model incorporating, among others, interpretations of character states, evolutionary mechanisms, various theoretical considerations, and ontogenetic and phylogenetic time.

An important part of the paper is devoted to a historical study of the type-concept, particularly in biology, in order to trace the origins of the concept, and to compare various methods and principles of type-construction and typology. This study produced several interesting results, among which not only a new view with reference to the origin of the eighteenth-century type-concept, but also the discovery of the German philosopher Teichmüller, unknown in biology, who in 1877 criticized Darwinism, and arrived by simple logical reasoning at the assumption of evolutionary mechanisms foreshadowing those described in my papers on Numerical Changes (Van der Hammen, 1981a) and Unfoldment and Manifestation (Van der Hammen, 1983).

In the last chapter of the present paper, all aspects of a type (an archetype) intended to serve as a model of evolutionary potentialities and a standard of higher classification will be further developed; these aspects are illustrated by examples from Chelicerata and other Arthropods. It may be remarked here that the section on the evolution of Cryptognomic characters, in the first part of my Comparative Studies in Chelicerata (Van der Hammen, 1979), constitutes a first outline of an archetype in the sense as further developed here.

2. Outline of the history of the type-concept, with particular reference to biology

Before starting on the development of a type-concept in the sense of norm or standard for the study of higher classification and model of evolution, it is necessary to make an introductory study of the development of the concept, particularly in biology, from Greek Antiquity up to the present. This study is divided into a number of subsections, successively devoted to: the period before the eighteenth century; Goethe's predecessors and contemporaries; systematic zoology; Goethe; Richard Owen; Charles Darwin and evolutionary biology; and the twentieth century. At the end of the section, these data are summarized and some general conclusions are drawn.

2.1. *The concepts of type and archetype before the eighteenth century*

Prior to a discussion of the type-concept in biology, from the eighteenth century to the present day, we shall briefly deal here with the use of type and archetype in philosophical literature of previous centuries. The following survey represents an outline of the subject, and does not constitute a complete history of these terms.

The word 'type' is derived from the Greek noun *typos*, which originally referred to a mould (a hollow form or matrix). As demonstrated by Von Blumenthal (1928), the current derivation, according to which the original meaning would refer to the impression made by a blow, is incorrect. The term was used by Plato (427–347 B.C.) in three different senses: (1) impression (*The Republic* II, 377b; *Theaetetus* 194b; *Timaeus* 71b); (2) model (*The Republic* II,

377c, 379a, 380c; III, 387c, 402d, 403e); (3) outline or survey (*Cratylus* 397a; *The Republic* III, 414a; VI, 491c; VIII, 559a). In the *Lexicon Philosophicum* by Johannes Micraelius (1597–1658) the term *typus* is defined as: (1) the original model of which any resemblance is made (exemplar, ad quod aliud exprimitur); and (2) an example signifying something beforehand, i.e. a symbol (exemplum aliquid praesignificans).

The word 'archetype' was used, about the beginning of our era, in Latin (archetypus), as a name for an example or the original. It is interesting to mention here that Aristotle (384–322 B.C.), in *The Metaphysics* (V, i), enumerated several meanings of *arche* (e.g. beginning, starting-point, foundation, origin, cause, directing principle, ruler), the common property of these being 'to be the first thing from which something either exists, or comes into being, or becomes known'; some of these beginnings 'are originally inherent in things, while others are not'. 'Archetype' was frequently used in Greek (archetypos) by Philo Judaeus (first century), in the metaphysical sense of Idea, i.e. the original (in the mind of God) of which all things are copies (*De Opificio Mundi* 16; *Legum Allegoria* II, 4; III, 96; *De Cherubim* 97; *Quod Deterius Potiori Insidiari Soleat* 83; *De Plantatione* 20, 50; *Quis Rerum Divinarum Heres* 225, 230; *De Somniis* I, 37, 75, 115, 173, 206, 232; *De Mutatione Nominum* 135, 183; *De Specialibus Legibus* I, 279; III, 83; *De Praemiis et Poenis* 163). It was used in a more or less similar way by several church-fathers and by Plotinus (205–270) (*Enneads* I, ii, 2; II, iv, 15; vi, 3; III, ii, 1; viii, 11; IV, iii, 13; V, i, 6; vii, 1; VI, ii, 7; iv, 10). The term was used by Giovanni Pico della Mirandola (1463–1494), in *De Dignitate Hominis*, in the sense of the original used by the Creator during the process of creation (Verum nec erat in archetypis unde novam sobolam effingeret); Pico influenced the Cambridge Platonists mentioned below.

Micraelius, in his above-mentioned Lexicon, defined 'archetype' in the sense of prototype or original model (Archetypa, prototypa, dicuntur causae exemplares in conceptu seu entellectu, cum quo res formandae debent congruere).

The term was used in England, in the course of the seventeenth century, in the circle of the Cambridge Platonists, for instance by Henry More (1614–1687), in the sense of Platonic Idea.[1]

Descartes (1596–1650) incidentally used the term archetype in his *Meditationes Prima Philosophia* (III, 43), first published in 1641. In the paragraph in question he explained that, in a series of ideas proceeding from each other, one arrives at last at an archetype or original idea (Et quamvis forte una idea ex alia nasci possit, non tamen hic datur progressus in infinitum, sed tandem ad aliquam primam debet deveniri, cujus causa sit instar archetypi, in quo omnis realitas formaliter contineatur, quae est in idea tantum objective). The term

---

[1] According to Murray (*A New English Dictionary*) the term 'archetypal' is found in More's *Psychozoia Platonica: or A Platonicall Song of the Soul* (note 146/1). The note is omitted in the edition consulted by me.

106

'archetype' was translated by 'un patron ou un original' in the French edition of the work, first published in 1647 (*Méditations* III, 44). It may be remarked here that to Descartes the term 'idea' had no Platonic connotation. In a letter to Hobbes, quoted by Fraser in his edition of Locke's *Essay Concerning Human Understanding* (vol. 1, p. lixn), Descartes wrote: 'je prends le nom d'idée pour tout ce qui est conçu immédiatement par l'esprit'.

The term archetype was used several times by John Locke (1632–1704), in his *Essay Concerning Human Understanding* (first published in 1690), particularly in book II (chapters 30–32) and book IV (chapter 4). The term pattern (in the sense of model) was used by him as a synonym.[2] Locke further developed Descartes' archetype-concept. According to him an archetype is a reality independent of the mind, a fixed standard external to our transitory ideas. Real ideas have a conformity with their archetypes. Moral as well as mathematical ideas are archetypes themselves.

Joseph Glanville (1636–1680), an English natural philosopher, who tried to find an empirical ground for the belief in the supernatural, used the archetype-concept, in his *Scepsis Scientifica* (1665), in the sense of: original of which any resemblance is made ('So our Souls, though they might have perceived the motions and images themselves by simple sense; yet without some implicit inference it seems inconceivable, how by that means they should apprehend their Archetypes').

Considering the current misconceptions with reference to the relation between typology and the Ideas in Platonic philosophy, the following note might be of interest here. The Greek words *eidos* and *idea* are derived from the same Indo-Germanic root with the meaning 'to see'. Both words originally referred to visible, external form. The Greek word *morphe* was also used with reference to form. Plato used several words instead of one, i.e. various expressions, for that conception which we now call a Platonic Idea (an eternally existing pattern of which the individual things are imperfect copies); the word *eidos* could form part of these expressions. Aristotle used the word *eidos* in opposition to matter. According to him, matter is that component of the essence of any thing, which requires the addition of a particular *eidos* to constitute it as determinately existent (cf. Bormann, Franzen, Krapiec & Oeing-Hanhoff, 1972; Meinhardt, 1972).

It may be added here that the term entelechy (entelecheia = perfection, realization) is connected with Aristotle's theory of *energeia* and *dynamis* (the term was probably coined by him). It is the realization of a potentiality (cf. Franzen, Georgulis & Nobis, 1972).

2. *The type-concept in literature by Goethe's predecessors and contemporaries*

In vol. 4 of *Histoire Naturelle, Générale et Particulière* (Buffon & Daubenton,

---

[2] In book III (chapter 3) the term 'form' is used as a synonym of 'type'.

1753), Buffon (1707–1788) used the type-concept in two related senses. On pp. 215–216, in the chapter on the horse, he used the term prototype in the sense of original (the first horse created) of which all individuals are copies (notwithstanding the considerable individual variation). Buffon's prototype is related to our genotype because it includes the complete phenotypic variation; it includes, moreover, the complete intraspecific (racial) variation. On pp. 379–381, in the chapter on the ass, Buffon used the type-concept (without mentioning the word), in a supraspecific connection, in the sense of original plan or model (dessein primitif et général) on which the designs of all animals are based. According to him, this original model appears to indicate that the Supreme Being, in creating the animals, used but a single Idea which he varied in all possible ways.[3] Buffon explains that the differences between two species (such as Man and Horse) are mainly differences in measurements, relations and numbers. One arrives at the concept of a single model by selecting one species as a base, to which the other species are compared. In this connection mention must be made also of Buffon's 'moules intérieures' or internal moulds, dealt with in vol. 2 of the same work (Buffon, 1749: 41–46). These moulds control the form of animals and plants during nutrition, development and reproduction (they could be compared with our genotype). It seems that Buffon's 'prototype', 'dessein primitif et général' and 'moules intérieures' are merged in the type-concept of the later eighteenth century.

In his *Pensées sur l'Interpretation de la Nature*, Denis Diderot (1713–1784) suggested that there could be one prototype for the two realms of nature, i.e. for all living beings (Diderot, 1754: 32–37 (Pensée XII); cf. Harris, 1981: 115). Diderot distinguished between the prototype and its envelope (which is the result of metamorphosis); this distinction still reminds us of the ancient theory of form and matter.

Jean-Baptiste-René Robinet (1735–1820), in his books on the great Chain of Being (Robinet, 1761–1766, 1768), also dealt with the type-concept. His views are discussed in detail by Thienemann (1909: 255–258) and Lovejoy (1936; pp. 269–283 in the edition of 1978). Robinet was influenced by Gottfried Wilhelm Leibniz (1646–1716), who is repeatedly quoted by him. He denied the existence of archetypes (Robinet, 1766: 'Dieu n'est point l'archétype du monde'. 'Dieu n'a besoin ni d'idée, ni d'archétype pour créer le monde'. 'Il n'y a point d'idée archétype; ces deux mots sont contradictoire'), but recognized a single plan of organisation (or animality) in nature, and a single prototypal Being. All differences in nature are natural variations of this prototype which must be regarded as 'l'élément générateur de tous les Êtres'. The fundamental element in nature is an organ in the shape of a hollow cylinder, com-

---

[3] Wilkie (1957: 49) quoted the following observation by Buffon (reported by Hérault de Séchelles, one of his collaborators): 'J'ai toujours nommé le créateur, mais il n'y a qu'à ôter ce mot, et mettre naturellement à la place la puissance de la nature, qui résulte des deux grandes lois, l'attraction et l'impulsion.'

posed, in its turn, of similar but much smaller organs (Robinet, 1766a). Robinet's prototype-concept is further developed in his book on the natural gradation of forms (Robinet, 1768). According to him, a prototype is an intellectual principle which realizes itself in matter (this realization being accompanied by change, i.e. deviation from the prototype); all variations in nature (a stone, an oak, a horse, an ape, man) are metamorphoses of the prototype. Man is the most perfect realization of the prototype (the prototype is man minus his physical realization). This seems to be a Scholastic view, but Robinet adopted it from Leibniz, who regarded a living animal as consisting of a substantial form (a substance of the nature of a soul) animating a bodily machine (Leibniz postulated a substantial form for each living organism; he was influenced in this by Scholasticism and Aristotle; cf. Broad, 1979: 75–81; Leibniz, *Philosoph. Schr.*, 2: 74–75; *Lettr. à Arnauld*: 52–54) (cf. essay I).

The German philosopher Johann Gottfried von Herder (1744–1803), in the first volume of his *Ideen zur Philosophie der Geschicht der Menschheit* (Herder, 1784), dealt with the fundamental identity of form in the animal kingdom (particularly in book 2 chapter 4, and book 5 chapter 1). He recognized a single principal form (Hauptform) or prototype, which manifests itself more and more clearly in an ascending series of forms. A detailed analysis of Herder's philosophy was published by Rouché (1940). Herder's philosophical and biological ideas were influenced by Leibniz, Buffon and Robinet. He had much influence on the morphological ideas of Goethe.

Félix Vicq d'Azyr (1748–1794) was the first to distinguish two different type-concepts, one with reference to comparative anatomy, a second with reference to various organs of one individual (i.e. to homonomous structures). Both concepts were discussed in his *Traité d'Anatomie et de Physiologie* (Vicq d'Azyr, 1786). Two passages from this treatise are quoted here. The first (p. 7), with reference to comparative anatomy, runs as follows: 'L'homme, isolé, ne paroît pas aussi grand, on ne voit pas aussi bien ce qu'il est: les animaux, sans l'homme, semblent être éloignés de leur type, et on ne sait à quel centre les rapporter'. The second passage (p. 11–12), with reference to the homonomous organs of one individual, runs as follows: 'L'anatomie comparée (...) n'est pas la seule à laquelle l'observateur puisse se livrer; il en est une autre qui mérite aussi son attention: son sujet (...) consiste dans l'examen des organes des mêmes individus comparés entr'eux. C'est aussi que les nerfs cervicaux peuvent être assimilés aux lombaires, les plexus axillaires aux sacrés (...). La Nature paroît donc suivre un type ou modèle général, non-seulement dans la structure des divers animaux, comme je l'ai déjà dit, mais encore dans celle de leurs différents organes'. The passages are repeated in the posthumous edition of the complete works (Vicq d'Azyr, 1805: 20–21, 31–32). The passage with reference to homonomous structures is formulated, in a different way, at another place in the same volume of the complete works (Vicq d'Azyr, 1805: 516): 'Et si les parties qui diffèrent le plus en apparence se ressembloient au fond, ne pourroit-on pas en conclure avec plus de certitude qu'il n'y a qu'un ensemble, qu'une forme essentielle, et que l'on reconnoit partout cette fécondité de la nature qui semble avoir imprimé à tous les êtres deux caractères nullement contradictoires, celui de la constance dans le type et de la variété dans les modifications?'. The last-mentioned formulation points already to the conception of a type composed of homonomous parts.

Latreille (1762–1833), in the third part of his *Histoire Naturelle Générale et Particulière des Crustacés et des Insectes* (Latreille, 1802), mentioned under each of his genera one or more species, indicated as examples ('exemples'). On p. 64 of this work he characterized one of his examples moreover as the insect which had served him in the formation of the genus. In vol. 7 of the same work (Latreille, 1804: 399) he used for the first time the word 'type' ('Le genre gamase a pour type le mite des Coléoptères de Geoffroy'). Finally, in his *Considérations Générales sur l'Ordre naturel des Animaux Composant les Classes des Crustacés, des Arachnides, et des Insectes* (Latreille, 1810: 421), he published a table of the genera with an indication of the species serving them as a type ('Table des genres avec l'indication de l'espèce qui leur sert de type'). Latreille regarded his *Histoire Naturelle* (on the title page and in the preface) as a continuation of Buffon's Natural History. Apparently, his type-concept was a further elaboration (at a lower level) of that introduced by Buffon, and mentioned at the beginning of this subsection (Buffon arrived at a model of a group by selecting one species as a base, to which other species could be compared; any species could be chosen for this purpose). Evidently, Latreille's types were selected as standards for comparison (a comparison enabling the conception of a model of the genus). He developed Buffon's concept in a particular way which now constitutes the base of type-species selection in modern systematic zoology.

Georges Cuvier (1769–1832), in his *Leçons d'Anatomie Comparée* (Cuvier, 1799: 58–60), distinguished between type and plan of construction. Animals more or less resembling each other can be arranged in a certain order, starting with the most perfect, and ending with the most simple. In this way they appear to be gradually removed from a primitive type (i.e. the most perfect animal, the one resembling most perfectly the original). Animals exhibiting a certain mutual resemblance appear to be constructed according to a common plan. Groups with different plans of construction are separated by discontinuities. It is, for instance, impossible to arrange Vertebrates and Invertebrates in a continuous linear series without important gaps. In the first two editions of his *Règne Animal*, Cuvier (1817: 57–60; 1829: 48–51) distinguished four different plans of animal construction (quatre formes principales, quatre plans généraux): Animalia vertebrata, Animalia mollusca, Animalia articulata, and Animalia radiata.

Étienne Geoffroy Saint-Hilaire (1772–1844) studied, in his *Philosophie Anatomique* (Geoffroy Saint-Hilaire, 1818), the unity of organic composition of the Vertebrates. He used the word 'type' sometimes as a synonym of plan of construction (p. xv), sometimes as a synonym of class (p. 3, 5). He studied

110

the analogy concept (which still included homology) and introduced the principle of connections (the constant position of an element among other elements; an organ can be reduced, but not transposed). The unity of composition
of animals was further discussed in his famous *Principes de Philosophie Zoologique* (Geoffroy Saint-Hilaire, 1830), the documents in the case of the Academy dispute. The word 'type' is used here in various ways. On p. 84 he mentions
the type of the bird (i.e. the model recognizable in all species). On p. 217−218
he mentions the 'ideal type' of the Vertebrate sternum. Whatever metamorphoses of this ideal type have taken place, it is not difficult 'd'en embrasser
tous les points communs, et de les ramener à une seule et même mesure, à des
fonctions identiques, enfin à un seul et même type'. In connection with the concept of metamorphosis, reference is made to Leibniz and his definition of the
universe as a unity in variety. Geoffroy Saint-Hilaire dismissed the suggestion
that his philosophy, influenced by the German natural philosophy, could be
characterized as pantheistic.

The French botanist Pierre-Jean-François Turpin (1775−1840) published an
*Essai d'une iconographie élémentaire et philosophique des végétaux, avec un
texte explicatif* (Turpin, 1820), which is accompanied by a folding plate, in the
text, pertaining to botanical organography. In the detailed explanation of the
plate, mention is made of the following: 'L'unité de composition organique est
le vœu de la nature. D'après cette grande loi on a pensé qu'avec la connaissance
de l'être vivant le plus compliqué, on pouvait parcourir comparativement l'en-
chainement gradué des Êtres, et se rendre facilement compte de toutes les
modifications qu'ils éprouvent en passant du plus simple au plus composé. Les
zoologistes trouvent dans l'homme, déjà très connu, un point de départ qui les
guide sûrement dans l'étude comparée des animaux et auquel ils peuvent tout
rapporter. En botanique, ce point de départ est encore à créer. (...) Dans ce
tableau d'organographie végétale, on s'est proposé deux choses: (1) de préciser
et de fixer les idées sur la connaissance des organes constitutifs pris isolément;
(2) de lier ces mêmes organes en raison de leur degré d'importance ou de l'ordre
qu'ils suivent dans leur développement. En mettant en rapport des Végétaux
et des Animaux, on ne les a considéré que dans leur vie organique, et on a seule-
ment voulu faire connaître, que les uns et les autres se composent de l'union
de deux systèmes d'organes, qui semblent, pour ainsi-dire, doubler la plupart
des êtres vivants.' In 1837, he prepared a synthetic drawing of an 'archetypal
plant' (Fig. 5.1), which was added (with an explanatory text) to a French translation of Goethe's works on natural history (Turpin, 1837). The drawing
represents a bare catalogue of the main varieties of form, known to him, arranged for the greater part without a theoretical base; it does not constitute a
scientific model.

Henri Ducrotay de Blainville (1777−1850) used the word 'type' in the sense
of large taxonomic group (a secondary division; the tertiary divisions were
called 'sous-type'). In 1816 Vertebrates and Invertebrates were called types, in
1840 Osteozoa (= Vertebrates), Entomozoa (= Arthropoda) and Malacozoa
(= Mollusca) (cf. Nicard, 1890: 63, 181).

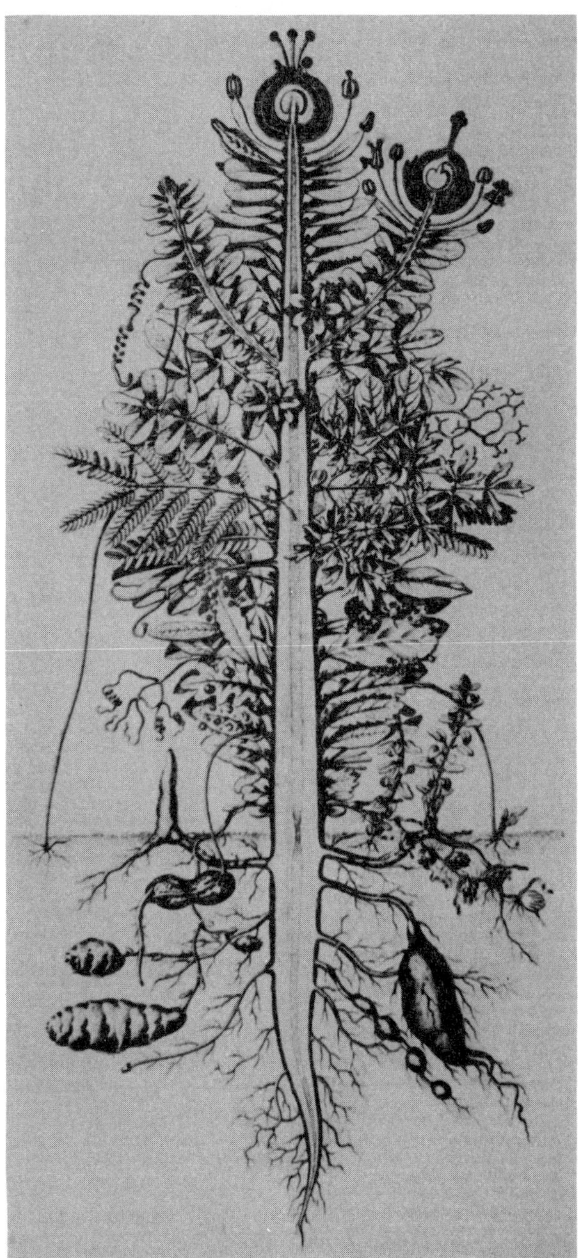

Fig. 5.1. Turpin's interpretation of Goethe's archetypcal plant, prepared for an outline of botanical organography, added to a French translation of Goethe's works on natural history. It is not a scientific model, but a schematic representation of a variety of organs arranged more or less arbitrarily along an axis. (After Turpin, 1837; reduced).

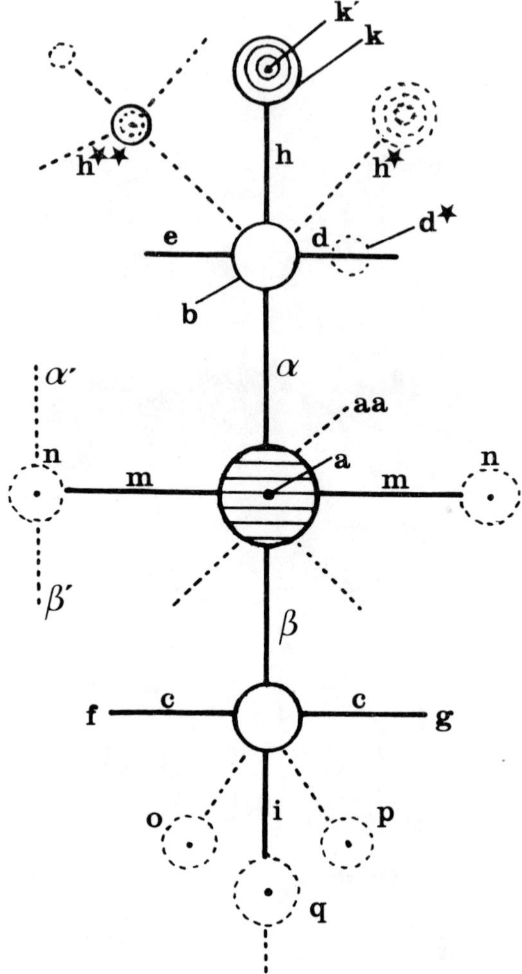

Fig. 5.2. Carus' interpretation of the archetypal plant; it constitutes an example of a diagrammatic, dynamic type. *a*: primordial cell; *as*: primordial node; *b*: upper secondary node (leaf-node); *c*: lower secondary node (root-node); *d, e*: opposite leaves; *d**, green 'leaf-tuber' (probably a bulbil) (imperfect repetition of the primordial cell); *f, g*: root fibrils (divisions of the lower secondary node); *h*: shoot of the stem, continued upwards; *h*, h***: potential lateral multiplication of the stem; *i*: shoot continued downwards (large tapering root); *k*: upper tertiary node (flower-bud); *k '*: primordial cell, produced by the flower-bud, after repeated threefold internal division (the same can be repeated at *h** and *h***); *mm*: horizontal shoot of the primordial node (offshoot) with new primordial node; *nn*: new primordial node; *o, p, q*: lower simple or compound nodes (root-tubers; *o* and *p* on lateral roots), that once again repeat (imperfectly) the primordial node; α, α' light-shoot; β, β', darkness-shoot. (Redrawn after Carus, 1861).

The physician, psychologist, philosopher and artist Carl Gustav Carus (1789–1869) dealt with the concept of the 'Urpflanze' (archetypal plant) and the concept of the 'Urwirbel' (archetypal vertebra, i.e. the archetypal element of the skeleton). His schematic representation of the archetypal plant (Fig. 5.2) is a diagrammatic, dynamic type. The archetypal element of the skeleton (i.e. the type of a series of homonomous elements) is schematically represented, in his book on Nature and Idea (Carus, 1861), by a ring-shaped segment.

In his *Beiträge zur Kenntnis der Niedern Thiere* (Von Baer, 1827: 731–762) and in the first part of his famous book on animal development (Von Baer, 1828: 206–231), the embryologist Karl Ernst von Baer (1792–1876) distinguished, in accordance with Cuvier, four main types of animal organization: the radiate type, the articulate type, the molluscan type, and the vertebrate type. He defined a type as the positional relations of the organic elements and organs. In a type, several grades of formation (characterized by a higher or lower degree of specialization and heterogeneity of the elements) can often be distinguished. The same grade of formation can be attained by different types. Von Baer (1828) studied also the embryological aspects of the type-concept. A type appears to control the embryonic development from the beginning (an embryo of a higher type does not pass through the lower types). In the course of embryonic development, the fundamental and more general characters are realized first, the more specialized characters later on. A main type is never exactly and completely realized, but always in a more or less modified way. Each main type is characterized by its own kind of development. A schematic representation of the embryonic development in a main type, demonstrates the genesis of the type. A type can be completely understood only from its way of development (Von Baer, 1828: 242–262).

2.3. *The type-concept in systematic zoology*

In the previous subsection mention is made of the introduction by Latreille, in systematic zoology, of the word 'type' (in the sense of type-species of a genus). This trend rapidly developed in the course of the nineteenth century. Von Heyden (1826), for instance, introduced a new classification of mites, in which a type is selected for each of his numerous genera. On the title-pages of the third edition of Cuvier's Règne Animal (Cuvier, 1836–1849), mention is made of plates representing the types of all genera (this type-concept is apparently based on that of Latreille, and not on that of Cuvier's *Leçons d'Anatomie Comparée*).

The further development of the type-concept in systematic zoology is closely connected with the development of rules for zoological nomenclature. This development dates back to a paper on zoological nomenclature published by Strickland (1837); in this paper type-species are not yet mentioned (although according to Strickland's rule no. 18 families and subfamilies should be based on the most typical genus). Probably on account of Strickland's paper, a committee was appointed by the British Association for the Advancement of Science, in 1842, to consider the rules by which the nomenclature of zoology might be established on a uniform and permanent basis. Strickland, Darwin and Owen were among the members of this Committee. It proceeded with great energy, and the rules could be published in 1843 (cf. Anonymous, 1843, 1843a, 1843b, 1843c). An Italian translation of the rules was adopted in Padua in 1843; the English edition was adopted in 1845 in America, and in 1846 in England by the British Association itself. Various editions, translations and reports were subsequently published, and a few slight changes were introduced (cf. Strickland, 1878). In these rules (the so-called Strickland Code) mention is made of type-species of genera (in the sense of standards or fixed points of reference), but not of type-specimens. On p. 7–8 of the 1878 edition, the type-species is connected with the typical portion of the original genus.

After 1850, some taxonomists (e.g. ornithologists) started using the word type also in the sense of type-specimen. In the Code of Nomenclature adopted by the American Ornithologists' Union, and first published in 1886 (cf. Anonymous, 1892: 52–53), rules are given for basing species and subspecies on a type-specimen. Suggestions for a more definite use of type-specimens were given by Thomas (1893). A summary of the rules of nomenclature, as used by entomologists at the end of the nineteenth century, was published by Walsingham & Durrant (1896); in this paper, type-specimens as well as types of supraspecific taxa are discussed.

Type-specimens are not mentioned in the rules adopted by the International Congresses of Zoology before 1913. Formal rules with reference to type-specimens were not adopted by the International Congresses before 1948.

2.4. *Goethe's type-concepts*

Johann Wolfgang von Goethe (1749–1832), the German poet, novelist, dramatist, statesman and scientist, occupies a special place in the history of the type-concept. He further developed the concept, and studied it in the fields of botany and zoology; his influence was far-reaching.

There are many editions of Goethe's literary and scientific works, and there are numerous studies on various aspects of his life and activities. For the present study I have made use of the modern standard edition of his scientific works, the so-called Leopoldina Ausgabe (abbreviated 'LA' when quoted below), in which the works with reference to botany and zoology are found in series 1 vols. 9 and 10, and series 2 vols. 9A and 9B (the second series, containing additional and explanatory material, is not yet complete). For his literary works, his letters, diaries, etc., I have made use of the so-called Artemis Gedenkausgabe (abbreviated 'AG' when quoted below). Goethe's drawings are published in the ten volumes of the *Corpus der Goethezeichnungen* (abbreviated 'Corpus' when referred to below); the zoological and botanical drawings are collected in vol. 5B. Bibliographical details of these editions are mentioned below, under the References. A bibliography of Goethe's work on the natural sciences (editions and studies) was published by Schmid (1940). Of the older studies on Goethe's biology and type-concepts, mention is made here of Bliedner (1901), Hansen (1919), Haecker (1927) and Spinner (1933). Of the more recent studies, mention must be made of Arber (1946), Wilkinson (1953), Bräuning-Oktavio (1956), Roger (1965), Nisbet (1967), Eyde (1975) and Wells (1978).

Goethe's philosophy was originally influenced in particular by Giordano Bruno (1548–1600) and Benedictus de Spinoza (1632–1677). In the dialogue *De la Causa, Principio et Uno* (written and first published in 1584), Bruno dealt with the problem of form and matter. According to him, form is immanent in matter (form is an aspect of matter); matter, form, potentiality and action are coincident. In proposition XL, note II of *The Ethics*, Part II (On the nature and origin of the mind) (The Ethics was first published in 1677), Spinoza wrote that there is a third kind of knowledge (besides opinion and reason), called intuition. This kind of knowledge proceeds from an adequate idea of the absolute essence of certain attributes of God to the adequate knowledge of the essence of things. This passage was quoted and commented upon by Goethe in a letter to Friedrich Heinrich Jacobi (dated 5 May 1786): '(...) ich halte viel aufs schauen, und wenn Spinoza von der Scientia intuitiva spricht, und sagt: Hoc cognoscendi genus procedit ab adaequata idea essentiae formalis quorundam Dei attributorum ad adaequatam cognitionem essentiae rerum; so geben mir diese wenigen Worte Muth, mein ganzes Leben der Betrachtung der Dinge zu widmen die ich reichen und von deren essentia formali ich mir eine adäquate Idee zu bilden hoffen kann (...)'. (AG, 18: 924).

Bräuning-Oktavio (1956) has demonstrated that Goethe did not start his own studies of the type-concept before 1785, and that his study of the intermaxillary bone, in 1784, was not connected with his later typological views. In 1786 Goethe resumed his botanical studies, and read Vicq d'Azyr's *Traité d'Anatomie et de Physiologie*. Shortly before his first journey into Italy, he developed his first ideas concerning an archetypal plant, and these ideas matured during this journey. Goethe's frequent discussions with Herder in 1783 and 1784, when the latter prepared his above-mentioned Ideas on the Philosophy of the History of Mankind, appear to have been of paramount importance to the subsequent development of his own type-concept. The development of his views, from the vague concept of some archetypal plant (influenced by Herder) to a pattern consisting of similar (homonomous) elements (as in Vicq d'Azyr's animal type) can be followed in his letters, diaries, Italienische Reise and botanical notes.

Besides 'Urpflanze' (AG, 11: 242, 291; 19: 84; LA(2), 9A: 357, 359, 365). Goethe used also the following terms to indicate the archetypal plant: 'Modell' (AG, 19: 84; LA(2), 9A: 365), 'Harmonia plantarum' (AG, 19: 93; LA(2), 9A: 370; the term is probably connected with Leibniz's 'Harmonia praestabilita' or pre-established harmony), ἐν καί πᾶν (AG, 11: 435; LA(2), 9A: 371; the expression, with the meaning 'the One and the All', is a quotation from Xenophanes), 'Formel' (formula; AG, 19: 95; LA(2), 9A: 373), 'Typus' (in notes from the period 1787–1789; LA(1), 10: 50; (2), 9A: 526, 533).

Goethe has left no complete drawing of the 'Urpflanze'. Several of his sketches (e.g. Corpus, 5B: figs. 90, 95) pertain to the Metamorphosis of Plants; one sketch (Corpus, 5B: Fig. 86) is a partial representation of the type of the annual higher plants, including the principles of metamorphosis. After Goethe, several authors have prepared schematic representations of the archetypal plant; those by Turpin, Schleiden, Unger, Sachs, Kerner von Marilaun, and Steiner were reproduced by Schmid (1930).

The development of Goethe's views from 1786, shortly before his journey into Italy, till the publication of the *Versuch die Metamorphose der Pflanzen zu erklären* (An attempt to interpret the metamorphosis of plants) in 1790, can be illustrated by the following quotations.

'Das Pflanzenreich rasst einmal wieder in meinem Gemüthe, ich kann es nicht einen Augenblick loswerden, mache aber auch schöne Fortschritte. (...) Am meisten freut mich ietzo das Pflanzenwesen, das mich verfolgt; und das ists recht wie einem eine Sache zu eigen wird. Es zwingt sich mir alles auf, ich sinne nicht mehr drüber, es kommt mir alles entgegen und das ungeheure Reich simplificirt sich mir in der Seele, dass ich bald die schwerste Aufgabe gleich weglesen kann. (...) Und es ist kein Traum oder Phantasie; es ist ein Gewahrwerden der wesentlichen Form, mit der die Natur gleichsam nur immer spielt und spielend das mannigfaltige Leben hervorbringt. Hätt ich Zeit in dem kurzen Lebensraum; so getraut ich mich es auf alle Reiche der Natur – auf ihr ganzes Reich – auszudehnen.' (Weimar, 9 July 1786, letter to Charlotte von Stein; AG, 18: 936–938; LA(2), 9A: 335–336).

'Schöne Bestätigungen meiner botanischen Ideen hab ich wieder gefunden. Es wird gewiss kommen und ich dringe noch weiter.' (Padua, 27 September 1786, diary entry; AG (suppl.), 2: 155; LA(2), 9A: 343).

'Hier in dieser neu mir entgegen tretenden Mannigfaltigkeit wird jener Gedanke immer lebendiger: dass man sich alle Pflanzengestalten vielleicht aus einer entwicklen könne. Hiedurch würde es allein möglich werden, Geschlechter und Arten wahrhaft zu bestimmen, welches, wie mich dünkt bisher, sehr willkürlich geschieht. Auf diesem Punkte bin ich in meiner botanischen Philosophie stecken geblieben und ich sehe noch nicht, wie ich mich entwirren will. Die Tiefe und Breite dieses Geschäfts scheint mir völlig gleich.' (Padua, 27 September 1786, Italienische Reise; AG, 11: 65; LA(2), 9A: 344).

'Eine Fächerpalme zog meine ganze Aufmerksamkeit auf sich; glücklicherweise standen die einfachen, lanzenförmigen ersten Blätter noch am Boden, die sukzessive Trennung derselben nahm zu, bis endlich das Fächerartige in volkommener Ausbildung zu sehen war. Aus einer spathagleichen Scheide zuletzt trat ein Zweiglein mit Blüten hervor, und erschien als ein sonderbares, mit dem vorhergehenden Wachstum in keinem Verhältnis stehendes Erzeugnis, fremdartig und überraschend.

'Auf mein Ersuchen schnitt mir der Gärtner die Stufenfolge dieser Veränderungen sämtlich ab, und ich belastete mich mit einigen grossen Pappen, um dieser Fund mit mir zu führen. Sie liegen, wie ich sie damals mitgenommen, noch wohlbehalten vor mir und ich verehre sie als Fetische, die, meine Aufmerksamkeit zu erregen und zu fesseln völlig geeignet, mir eine gedeihliche Folge meiner Bemühungen zuzusagen schienen.'

'Das Wechselhafte der Pflanzengestalten (. . .) erweckte nun bei mir immermehr die Vorstellung: die uns umgebenden Pflanzenformen seien nicht ursprünglich determiniert und festgestellt, ihnen sei vielmehr, bei einer eigensinnigen, generischen und spezifischen Hartnäckigkeit, eine glückliche Mobilität und Biegsamkeit verliehen, um in so vielen Bedingungen, die über dem Erdkreis auf sie einwirken, sich zu fügen und darnach bilden und umbilden zu können.' (Padua, 27 September 1786; Der Verfasser teilt die Geschichte seiner botanischen Studien mit, 1831; LA(1), 10: 333).

'Meine botanischen Grillen bekräftigen sich an allem diesen, und ich bin auf dem Wege, neue schöne Verhältnisse zu entdecken, wie die Natur, solch ein Ungeheures, das wie nichts aussieht, aus dem Einfachen das Mannigfaltigste entwickelt.' (Rome, 19 February 1787, Italienische Reise; AG, 11: 190; LA(2), 9A: 353).

'Herdern bitte zu melden, dass meine botanischen Aufklärungen weiter und weiter gehen; es ist immer dasselbe Prinzip, aber es gehörte ein Leben dazu um es durchzuführen. Vielleicht bin ich noch imstande die Hauptlinien zu ziehen.' (Naples, 13 March 1787, Italienische Reise; AG, 11: 224; LA(2), 9A: 357).

'Da kam mir eine gute Erleuchtung über botanische Gegenstände. Herdern bitte ich zu sagen, dass ich mit der Urpflanze balde zu Stande bin, nur fürchte ich, dass niemand die übrige Pflanzenwelt darin wird erkennen wollen. Meine famose Lehre von den Kotyledonen ist so sublimiert, dass man schwerlich wird weiter gehen können.' (Naples, 25 March 1787, Italienische Reise; AG, 11: 242−243; LA(2), 9A: 357).

'Heute früh ging ich (. . .) nach dem öffentlichen Garten. (. . .) Im Angesicht so vielerlei neuen und erneuten Gebilden fiel mir die alte Grille wieder ein: ob ich nicht unter dieser Schar die Urpflanze entdecken könnte? Eine solche muss es denn doch geben! Woran würde ich sonst erkennen, dass dieses oder jenes Gebilde eine Pflanze sei, wenn sie nicht all nach einem Muster gebildet wären.'

'Ich bemühte mich zu untersuchen, worin denn die vielen abweichenden Gestalten voneinander unterschieden seien. Und ich fand sie immer mehr ähnlich als verschieden, und wollte ich meine botanische Terminologie anbringen, so ging das wohl, aber es fruchtete nicht, es machte mich unruhig, ohne dass es mir weiter half.' (Palermo, 17 April 1787, Italienische Reise; AG, 11: 291; LA(2), 9A: 359).

'Wie sie sich nun unter einen Begriff sammeln lassen, so wurde mir nach und nach klar und klärer, dass die Anschauung noch auf eine höhere Weise belebt werden könnte: eine Forderung, die mir damals unter der sinnlichen Form einer übersinnlichen Urpflanze vorschwebte. Ich ging allen Gestalten, wie sie

118

mir vorkamen, in ihren Veränderungen nach, und so leuchtete mir am letzten
Ziel meiner Reise, in Sizilien, die ursprüngliche Identität aller Pflanzenteile
volkommen ein, und ich suchte diese nunmehr überall zu verfolgen und wieder
gewahr zu werden.' (Metamorphose 1831; LA(1), 10: 334).

'Sage Herdern dass ich dem Geheimnis der Pflanzenzeugung und Organisation ganz nah bin und dass es das einfachste ist was nur gedacht werden kann.
Unter diesem Himmel kann man die schönsten Beobachtungen machen. Sage
ihm dass ich den Hauptpunkt wo der Keim stickt ganz klar und zweifellos ent-
deckt habe, dass ich alles übrige schon im ganzen übersehe und nur noch einige
Punkte bestimmter werden müssen. Die Urpflanze wird das wunderlichste
Geschöpf von der Welt über welches mich die Natur selbst beneiden soll. Mit
diesem Modell und dem Schlüssel dazu, kann man alsdann noch Pflanzen ins
Unendliche erfinden, die konsequent sein müssen, das heisst: die, wenn sie
auch nicht existieren, doch existieren könnten und nicht etwa malerische oder
dichterische Schatten und Scheine sind, sondern eine innerliche Wahrheit und
Notwendigkeit haben. Dasselbe Gesetz wird sich auf alles übrige Lebendige an-
wenden lassen.' (Rome, 8 June 1787, letter to Charlotte von Stein; AG, 19:
84−85; LA(2), 9A: 365).

'Es war mir nämlich aufgegangen dass in demjenigen Organ der Pflanze,
welches wir als Blatt gewöhnlich anzusprechen pflegen, der wahre Proteus ver-
borgen liege, der sich in allen Gestaltungen verstecken und offenbaren könne.
Vorwärts und rückwärts ist die Pflanze immer nur Blatt, mit dem künftigen
Keime so unzertrennlich vereint, dass man eins ohne das andere nicht denken
darf. Einen solchen Begriff zu fassen, zu ertragen, ihn in der Natur aufzufin-
den ist eine Aufgabe, die uns in einen peinlich süssen Zustand versetzt.' (Rome,
July 1787, Italienische Reise; AG, 11: 413−414; LA(2), 9A: 366).

'Ich hoffe du wirst auch dereinst an meiner Harmonia Plantarum, wodurch
das Linnéische System aufs schönste erleuchtet wird, alle Streitigkeiten über
die Form der Pflanzen aufgelöst, ja sogar alle Monstra erklärt werden Freude
haben.' (Rome, 18 August 1787, letter to Karl von Knebel; AG, 19: 93; LA(2),
9A: 370).

'In der Naturgeschichte bring ich dir Sachen mit, die du nicht erwartest. Ich
glaube dem Wie der Organisation sehr nahe zu rücken. Du sollst diese Manifes-
tationen (nicht Fulgurationen) unsres Gottes mit Freuden beschauen und mich
belehren, wer in der alten und neuen Zeit dasselbe gefunden gedacht, es von
eben der Seite oder aus einem wenig abweichenden Standpunkte betrachtet.'
(Rome, 28 August 1787, Italienische Reise; AG, 11: 428−429; LA(2), 9A: 371).

'Mich hat er aufgemuntert in natürlichen Dingen weiter vorzudringen, wo
ich denn besonders in der Botanik auf ein ἐν καί πᾶν gekommen bin, das mich
in Erstaunen setzt; wie weit es um sich greift, kann ich selbst nicht sehn.'
(Rome, 6 September 1787, Italienische Reise; AG, 11: 435; LA(2), 9A: 371).

'Ich versuchte es mit Moritz und trug ihm, so viel ich vermochte, die
Metamorphose der Pflanzen vor; (. . .).' (Frascati, 28 September 1787, Italienische Reise; AG, 11: 446; LA(2), 9A: 373).

'Die Botanik übe ich auf Wegen und Stegen. Es möchte wie eine Rodomontade klingen, wenn ich sagte, wie weit ich darin gekommen zu sein glaube. Genug ich werde immer sichrer dass die allgemeine Formel die ich gefunden habe, auf alle Pflanzen anwendbar ist. Ich kann schon die eigensinnigsten Formen, zum Exempel Passiflora, Arum, dadurch erklären und miteinander in Parallel setzen. Zur völligen Ausbildung dieser Idee brauchts doch noch Zeit. Dieses Land ist schon recht zu einem solchen Studio gemacht. Was ich im Norden nur vermutete finde ich hier offenbar. Leider dass ich so ganz von allen Büchern, die zu diesem Studio gehören, entfernt bin!' (Frascati, 3 October 1787, letter to Karl von Knebel; AG, 19: 95; LA(2), 9A: 373).

'Indessen bin ich auch angespornt worden meine botanischen Ideen zu schreiben.' (20 November 1789, letter to Herzog Karl August von Sachsen-Weimar; LA (2), 9A: 386).

'Auf der Rückreise verfolgte ich unablässig diese Gedanken, ich ordnete mir im stillen Sinne einen annehmlichen Vortrag dieser meiner Ansichten, schrieb ihn bald nach meiner Rückkehr nieder und liess ihn drucken. Er kam 1790 heraus und ich hatte die Absicht bald eine weitere Erläuterung mit den nötigen Abbildungen nachfolgen zu lassen.' (Metamorphose 1831; LA(1), 10: 336).

'Nach Deutschland endlich zurück getrieben (...) fühlte ich Wert und Würde des Naturelements desto lebhafter. Da suchte ich Heil und Behagen, ergriff mit Leidenschaft alle frühere Fäden, die mich an Naturforschung und ihre Freunde knüpfen sollten, und eine meiner ersten Arbeiten war der Aufsatz, welcher nunmehr abermals abgedruckt, sich als bekannt und, nach fast dreissig Jahren endlich im Kreise der Wissenschaft aufgenommen, Freunde und Bemerkern lebendiger Naturwirkungen zu Gunst und Huld aufs neue gar zu gern empfehlen möchte.' (Entstehen des Aufsatzes über Metamorphose der Pflanzen 1817; LA(1), 9: 21−22).

In the *Versuch die Metamorphose der Pflanzen zu erklären*, published in 1790, no mention is made of the 'Urpflanze'. The essay deals with the fundamental similarity of the appendages of the stem (seed-leaves, stem-leaves, calyx, corolla, androecium, nectaries, style, fruit). It was but part of Goethe's botanical philosophy, and he indeed made many notes for a continuation. It has been suggested that Goethe had abandoned his earlier views on the archetypal plant, and up to now Goethe's essay has generally not been connected with Vicq d'Azyr's views on the composition of the animal type of similar parts (later on called 'homotypes' by Owen).[4] Goethe had indeed read Vicq d'Azyr in 1786, shortly before the rapid development of his botanical views. He was probably not conscious of the influence. Caspar Friedrich Wolff (1733−1794)

---

[4] As mentioned above, these homotypes can now be understood as developmental variations of a single genetic pattern. Metamorphosis (Goethe applied the term to his metamorphosis of plants as well as to the postembryonic development of insects) evidently pertains to this developmental variation.

who in 1768, more than twenty years before the publication of Goethe's essay, published a paper on the similarity of the appendages of the higher plant, did apparently not consider the similar parts in relation to a type (Goethe discovered this paper much later).

Goethe's archetypal plant must be regarded as a model related to Leibniz's substantial form; his first ideas must have been strongly influenced by Herder and, later on, by Vicq d'Azyr. This substantial form is immanent (in Bruno's philosophy form was an aspect of matter). Goethe scholars have generally not recognized this immanent character; mention must, however, be made of Steiner (1897; p. 22–23, 46–47 in the edition of 1963), Steiner (1926; p. 99 in the edition of 1973), and Ziehen (1930: 39, 46–47). The last-mentioned author characterized Goethe's botanical type as a 'Komplex von Bildungsgesetzen', and compared it with Aristotle's formative principles.

In order to understand the nature of Goethe's archetypal plant, we should compare it now to our concept of a genotypic pattern with its developmental and evolutionary potentialities. Goethe's model is a visual representation of this pattern, obtained by Spinoza's intuitive science.

Goethe had already studied anatomy in 1781, under the guidance of Justus Christian von Loder, professor of anatomy at the University of Jena. Prior to his study of the metamorphosis of plants he had studied the intermaxillary bone. After his return from Italy, in 1788, he occupied himself again with zoology. A chance discovery, during his short second journey into Italy, in 1790, led him to the application of his botanical theory to the study of mammals. The fact is that, in April 1790, he discovered, in the sand of the Jewish cemetary at Venice, a sheep's skull. Goethe suddenly realized (he later referred to this mental process as 'gegenständliches Denken') that the bones of the skull must be derived from vertebrae. The discovery is documented by two letters, from which I give the following two quotations.

'Sagen Sie Herdern dass ich der Tiergestalt und ihren mancherlei Umbildungen um eine ganze Formel näher gerückt bin und zwar durch den sonderbarsten Zufall.' (Venice, 30 April 1790, letter to Charlotte Sophie Juliane von Kalb; AG, 19: 64; LA(2), 9A: 392).

'Durch einen sonderbar glücklichen Zufall, dass Götze zum Scherz auf dem Judenkirchhof ein Stück Tierschädel aufhebt und ein Spässchen macht, als wenn er mir einen Judenkopf präsentierte, bin ich einen grossen Schritt in der Erklärung der Tierbildung vorwärts gekommen. Nun steh ich wieder vor einer andern Pforte, bis mir auch dazu das Glück den Schlüssel reicht.' (Venice, 4 May 1790, letter to Karoline Herder; AG, 19: 165; LA(2), 9A: 392).

Goethe did not publish his discovery until 1820, certainly because, as he mentioned on that occasion, it could not be proved (*Zur Morphologie*, 1(2); cf. LA(1), 9: 185). In the mean time Oken had also discovered the vertebral origin of the skull; he published his discovery in 1807. Goethe derived the mammal skull from six vertebrae, Oken from four. Goethe returned to the vertebral origin of the skull in *Zur Morphologie*, 2(1), published in 1823, and in *Zur Mor-*

phologie, 2(2), published in 1824 (cf. LA(1), 9: 309, 357–358).

Evidently, Goethe's theory of the vertebral origin of the skull is related to his essay on the metamorphosis of plants. Both deal with a type's composition of similar parts, i.e. with the repetition of similar elements in a plan of construction.

Although the theory of the vertebral origin of the skull has proved to be erroneous and is now abandoned, modern morphology has demonstrated the segmental origin of the region of the head (cf. Peyer, 1950).

In 1790, Goethe started the study of an osteological type. A manuscript of that year, entitled 'Versuch über die Gestalt der Tiere', contains his first ideas on that subject (LA(1), 10: 74–87; cf. LA(2), 9A: 564–571). Goethe makes here an attempt to draft a type which can serve as a guiding principle in the labyrinth of forms. For that purpose, a general scheme must be drawn up, to which the skeleton of man as well as that of mammals must be subordinated, and with which these can be compared. Goethe's problems concern the recognition of elements in the case of fusion, and the occurrence of different numbers of similar elements (homology problems). Besides that, he is dealing with the occurrence of similar elements in one individual (homonomy problems). The following quotation is interesting because it corresponds with the line of thought of Vicq d'Azyr mentioned above: 'Geben wir genau auf diese Mannigfaltigkeit acht so werden wir in der Stand gesetzt, nicht allein die Tiere unter einander sondern sogar das Tier mit sich selbst zu vergleichen. In dieser bei genauer Betrachtung die grösste Bewunderung erregende Veränderlichkeit der Teile, ruht die ganze Gewalt der bildenden Natur.' (LA(1), 10: 80).

In 1791, Goethe started his optical studies which took much of his time. It is generally assumed that the manuscript of the 'Versuch einer allgemeinen Knochenlehre' (LA(1), 10: 87–109; cf. LA(2), 9A: 572–575), in which the study of the osteological type is continued, dates from 1794, as do the 'Weitere Beschreibung zur Ergänzung der Knochenlehre' (LA(1), 10: 109–118; cf. LA(2), 9A: 576–579) and the 'Versuch einer allgemeinen Vergleichungslehre' (LA(1), 10: 118–122; cf. LA(2), 9A: 580–584). These manuscripts show that a change in Goethe's typological views and methods had set in. This change is documented by the following two quotations. The first pertains to a discussion with the German poet Friedrich von Schiller (1759–1805). 'Wir gelangten zu seinem Hause, (. . .) da trug ich die Metamorphose der Pflanzen lebhaft vor, und liess, mit manchen charakteristischen Federstrichen, eine symbolische Pflanze vor seinen Augen entstehen. Er vernahm und schaute das alles mit grosser Teilnahme, mit entschiedener Fassungskraft; als ich aber geendet, schüttelte er den Kopf und sagte: das ist keine Erfahrung, das ist eine Idee. (. . .) ich nahm mich aber zusammen und versetzte: das kann mir sehr lieb sein dass ich Ideen habe ohne es zu wissen, und sie sogar mit Augen sehe.' (21 July 1794; LA(1), 9: 79–83).

The second quotation refers to Goethe's Kant studies. It is significant that this passage, in *Zur Morphologie*, is inserted between the parts with reference to respectively the botanical and the osteological type (the autobiographical arrangement of *Zur Morphologie* is well-known). From discussions of Kant's *Kritik der Reinen Vernunft*, Goethe had understood 'wie viel unser Selbst und wie viel die Aussenwelt zu unserm geistigen Dasein beitrage. Ich hatte beide niemals gesondert, und wenn ich nach meiner Weise über Gegenstände philosophierte, so tat ich es mit unbewusster Naivetät und glaubte wirklich ich sähe meine Meinungen vor Augen. (. . .) Die Erkenntnisse a priori liess ich mir auch gefallen, so wie die synthetische Urteile a priori: denn hatte ich doch in meinem ganzem Leben, dichtend und beobachtend, synthetisch, und dann wieder analytisch verfahren, die Systole und Diastole des menschlichen Geistes war mir, wie ein zweites Atemholen, niemals getrennt, immer pulsierend.' (LA(1), 9: 91). Goethe was more enthusiastic in his remarks on Kant's *Kritik der Urteilskraft* (which first appeared in 1790). Evidently, he assimilated fragments of Kantian philosophy in a particular way. When he gave an exposition of his philosophy, orthodox Kantians used to remark 'es sei freilich ein Analogon Kantischer Vorstellungsart, aber ein seltsames.' (LA(1), 9: 92).

Kuhn (1967: 132–133) prepared a characteristic of Goethe's second typological method; it comprises the following steps: observation of the outer world; analysis of the facts with the help of the experience; synthesis by means of the imaginative faculty; testing, subjection to reasoning, and development of a concept or idea; renewed views of the outer world.

Goethe further developed this second typological approach in his 'Erster Entwurf einer allgemeinen Einleitung in die vergleichende Anatomie, ausgehend von der Osteologie' (manuscript dating from 1795, published in *Zur Morphologie*, 1(2) in 1820; cf. LA(1), 9: 119–151) and in his 'Vorträge über die drei ersten Kapitel des Entwurfs einer allgemeinen Einleitung in die vergleichende Anatomie, ausgehend von der Osteologie' (manuscript dating from 1796, published in *Zur Morphologie*, 1(3) in 1820; cf. LA(1), 9: 193–209). The following passages are quoted from the 'Erster Entwurf'.

'Deshalb geschieht hier ein Vorschlag zu einem anatomischen Typus, zu einem allgemeinen Bilde, worin die Gestalten sämtlicher Tiere, der Möglichkeit nach, enthalten waren, und wornach man jedes Tier in einer gewissen Ordnung beschriebe. Dieser Typus müsste so viel wie möglich in physiologischer Rücksicht aufgestellt sein. Schon aus der allgemeinen Idee eines Typus folgt, dass kein einzelnes Tier als ein solcher Vergleichungskanon aufgestellt werden könne; kein Einzelnes kann Muster des Ganzen sein.' (LA(1), 9: 121).

'Die Erfahrung muss uns vorerst die Teile lehren, die allen Tieren gemein sind, und worin diese Teile verschieden sind. Die Idee muss über dem Ganzen walten und auf eine genetische Weise das allgemeine Bild abziehen. Ist ein solcher Typus auch nur zum Versuch aufgestellt, so können wir die bisher gebräuchlichen Vergleichungsarten zur Prüfung desselben sehr wohl benutzen.' (LA(1), 9: 121).

'Nach aufgebautem Typus verfahrt man bei Vergleichung auf doppelte Weise. Erstlich dass man einzelne Tierarten nach demselben beschreibt. (. . .)

Sodann kann man aber auch einen besondern Teil durch alle Hauptgattungen durch beschreiben, wodurch eine belehrende Vergleichung volkommen bewirkt wird. (. . .) Doch müsste man vorerst über ein allgemeines Schema sich verständigen, worauf das Mechanische der Arbeit durch eine Tabelle befördert werden könnte (. . .).' (LA(1), 9: 122).

Goethe continues (in his 'Erster Entwurf') with a subdivision of the type into head, median part, and posterior part, and adds the following interesting remarks: '(. . .) dass keinem Teil etwas zugelegt werden könne, ohne dass einem andern dagegen etwas abgezogen werde, und umgekehrt.' (LA(1), 9: 124). '(. . .) das Tier wird durch Umstände zu Umständen gebildet; daher seine innere Volkommenheit und seine Zweckmässigkeit nach aussen.' (LA(1), 9: 126).

Goethe's first draft of the osteological type is an extensive scheme, elaborated in great detail. The use of terms like 'Idee' and 'Zweckmässigkeit' point to the Kantian influences.

In the last years of his life, Goethe showed his continued interest in typology by his notes on the above-mentioned Paris Academy dispute, between Cuvier and Geoffroy Saint-Hilaire (LA(1), 10: 373–403; cf. Kuhn, 1967).

2.5. *The type-concept of Richard Owen*

The development of the type-concept in comparative animal anatomy reached a pinnacle in the work of the English morphologist Richard Owen (1804–1892). Owen developed his type-concept at about the same time as his homology-concepts, and all these concepts are mutually closely connected. He distinguished three different homology-concepts: special, serial, and general homology. The first corresponds with the modern homology concept, the second and the third with the modern homonomy concept (serial homology refers to the similarity of metameres or metameric structures in one individual; general homology refers to the general similarity of any elements of one individual). The general type of the vertebrate skeleton, which illustrates the homologies in the group, is represented by a series of similar (homonomous) elements.

Owen was influenced, among others, by Vicq d'Azyr, Goethe, and Geoffroy Saint-Hilaire. According to Huxley (in: Owen, 1894, vol. 2: 312–318), he was moreover influenced by Oken (1779–1851) and probably also by Cudworth (1617–1688), one of the Cambridge Platonists. Owen wrote indeed an article on Oken in *The Encyclopaedia Britannica* (Owen, 1858), whilst the translation into English of Oken's *Lehrbuch der Naturphilosophie* (Oken, 1847) was undertaken at his instance.

Owen first dealt with the type-concept in his *Lectures on the Comparative Anatomy and Physiology of the Vertebrate Animals* (Owen, 1846: 41). In order to understand the fundamental type of the vertebrate skeleton, he started its study in the lowest class (the Fishes) where the primitive type is least obscured by adaptations. In order to detect the true general type, the study was extended ιo a comparison with the higher classes. The first step to this knowledge is the determination of the segments constituting the vertebrate skeleton. Owen applied the term 'vertebra' (in an extended sense) to each of these primary segments.

His type-concept was further developed in the two editions of his detailed study *On the Archetype and Homologies of the Vertebrate Skeleton* (Owen, 1847, 1848). In these publications the term 'type' is partly replaced by 'archetype'. Owen (1848: 177n) explains that the latter term is an equivalent of his 'general type' and 'fundamental type', mentioned in his 1846 book, and that he has used it in the sense which it bears in Joseph Glanvill's *Scepsis Scientifica* (1665), Isaac Watts's *Logick* (1725), and Samuel Johnson's dictionary.[5] Owen (1847, 1848) referred also to Maclise's *Comparative Osteology* (Maclise, 1847: introduction, p. 10; remarks on pls. 15, 16) in which the term archetype is used consistently. Maclise apparently based his archetype-concept on Goethe's type-concept, and used it in combination with the metamorphosis-concept (he knew Goethe's works on Natural History from a French translation). He defined 'archetype' as the unity upon which the law of form (metamorphosis) operates; the archetype is the complete form (cf. substantial form), whilst a vertebra stands as part of this structure (a vertebra is to be regarded as a proportional of some fuller archetype structure).

Owen (1848: 7) repeated that the general type of the vertebrate endo-skeleton is rightly represented by the idea of a series of essentially similar segments succeeding each other, and that any given part of one segment could be repeated in the rest of the series. Owen (1848: 175–177, Pl. 2 Fig. 1) prepared also a diagram of the ideal pattern or archetype of the vertebrate endoskeleton (cf. Fig. 5.3). On p. 175 of his book (Owen, 1848), the term 'homotype' is used in connection with a segment, Owen joined Goethe and Oken in accepting the vertebral theory of the skull; as mentioned above, this theory is incorrect, although modern morphology has demonstrated the segmented origin of the region of the head (Peyer, 1950: 50).

---

[5] The replacement of 'type' by 'archetype' could be connected with the introduction of 'type' in the rules for zoological nomenclature. As mentioned above, Owen was a member of the committee, appointed by the British Association for the Advancement of Science, to consider the rules by which zoological nomenclature might be established on a uniform and permanent basis. The rules were adopted in England in 1846.

Isaac Watts (1674–1748), a popular writer of hymns, educational manuals and philosophical books, mentioned the archetype-concept in his *Logick*, first published in 1725 ('As a man, a tree, are the outward objects of our perception, and the outward archetypes of our ideas; so our sensations of hunger, cold, are also inward archetypes or patterns of our ideas. But the notions or pictures of these things, as they are in the mind, are the ideas.'). Evidently, Watts' archetype-concept corresponds with that of John Locke, whilst that of Owen corresponds with the type-concept of Vicq d'Azyr and Goethe.

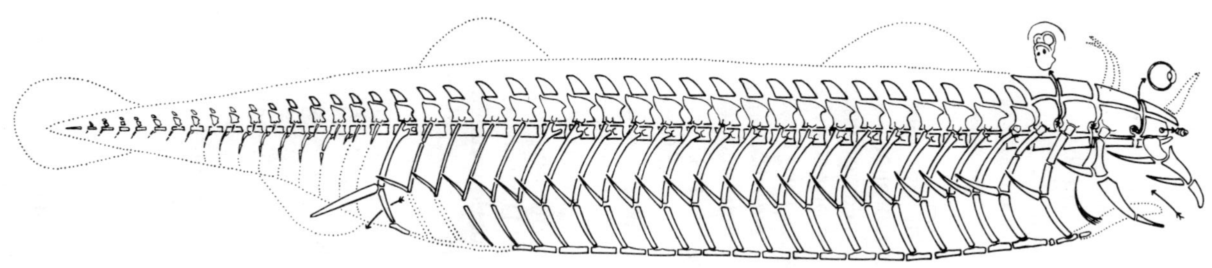

Fig. 5.3. Owen's ideal pattern or archetype of the vertebrate endoskeleton (lateral view). It is composed of a series of (originally homonomous) segments. The model contains a number of erroneous interpretations (among which those connected with the vertebral theory of the skull), but still constitutes an interesting construction. (Redrawn after Owen, 1848; reduced).

2.6. *Charles Darwin and the type-concept in evolutionary biology*

In Chapter 6 of *The Origin of Species*, Charles Darwin (1809–1882) explained unity of type by unity of descent. In Chapter 14, he used the word 'type' several times in the sense of plan of construction. He also dealt with 'serial homology' (homonomy) and 'metamorphosis', in the cases of leaves and the elements of the flower, and vertebrae and the skull; he introduced the hypothesis of their having been metamorphosed, not one from the other, as they now exist, but from some common and simpler element.

Darwin mentioned Buffon, Lamarck, Goethe, Geoffroy Saint-Hilaire and Owen among his predecessors. Several authors have discussed the presence of evolutionary ideas in the works of some of these (Wilkie, 1957, 1959; Michéa, 1958; Wells, 1978). The germs of a complete evolutionary theory were indeed present in the works of the typological morphologists. As will be demonstrated below, some of these germs have remained nearly undeveloped because of the exclusive consideration of the principle of natural selection.

The term 'archetypal' is mentioned in the glossary (prepared by Dallas) at the end of the sixth edition of *The Origin of Species* (Darwin, 1872), because the term is found in the Historical Sketch, in a sentence quoted from Owen.

The metaphysician Gustav Teichmüller (1832–1888), who was influenced by Leibniz, made a very important study of Darwinism and Philosophy (Teichmüller, 1877), in which much attention is paid to the type-concept, and in which he was far ahead of his time. The study remained practically unobserved. He defined a type as a normative form, and distinguished it from the form as it manifests itself. His types are, consequently, innate genotypic patterns underlying the external form. He criticized Darwinism because of its over-emphasizing of external causes. Transformation can only take place by a co-ordination of internal and external forces, although the laws of transformation (and the evolutionary potentialities) are within the type. A change in the internal system influences all parts of the type. The origin of new types consequently requires a co-ordinated change of the complete system; this change could be connected with the origin of new geological periods.

2.7. *The type-concept in twentieth century biology*

The type-concept was thoroughly studied by the Swiss morphologist Adolf Naef. In a publication on Idealistic Morphology and Phylogeny (Naef, 1919), he defined a type as a form present in mind as the result of thinking, the Idea of a natural being. The separate forms existing in nature are variations on this type; they are linked with it by series of transitional forms. A type must include an ontogenetic dimension (in which homologous stages can be established). Types can be arranged in the hierarchy of the natural system. They can be constructed by comparison of separate forms. Naef mentioned the following criteria and auxiliary criteria for the construction of types: more typical character states precede less typical character states in ontogeny and phylogeny; more typical character states are those mentioned for higher taxonomic categories, those correlated with other typical character states, those more fully developed, and those representing average values (the last-mentioned three are auxiliary criteria). Types could, in some cases, have existed as real ancestral forms. Phylogenetic methods should be based on idealistic morphology. Naef (1931) subsequently defined types as norms represented by illustrations (and in this way different from plans of construction and diagnoses; a type comprises all that is normal for a group; a plan of construction or a diagnosis comprise the common characters only). Types are based on empirical data (although they can be first understood intuitively). It is possible to deal with the types of separate structures (e.g. the type of a Crustacean walking-leg).

A detailed study of the type-concept and the construction of types was made by Smirnov (1925). According to him, a type must constitute a standard of comparison, and at the same time be characteristic of a whole group. Types are synthetical; they cannot be produced by abstraction (a diagnosis is an abstraction). They must be constructed from the average values of the different characters. Smirnov illustrated his method by the construction of types of the male genital system of groups of Syrphidae (Diptera).

According to Troll (1928), morphology is an idealistic science, related to the theory of Platonic Ideas. A type is an Idea which can only be understood intuitively. It manifests itself in separate organisms, where it is present as the plan of construction. The morphological connections between the members of a circle of forms are of a spiritual nature. Individual forms must be arranged in such a way that the type becomes visible behind them. A type is not the central value of a circle of forms (i.e. not static); it unfolds in space and time in a definite direction, more or less corresponding with the direction of evolution (it is dynamic). Troll's theoretical considerations appear to refer to a personal, extra-scientific philosophy; this does not alter the fact that his achievements in the field of botanical morphology are of paramount importance. Similar views are repeated in his collected essays on general morphology (Troll, 1941). In this book, he emphasized the impossibility of explaining unity of plan of construction in terms of modern genetics. He defined comparative morphology as a theory of type, i.e. typology. Troll exaggerated perhaps Goethe's importance in the foundation of typology; Goethe evidently continued the tradition of Buffon, Robinet and Vicq d'Azyr.

Van der Klaauw (1931) made a study of the norm-concept (which he characterized as imperfect, a concept in the making), and prepared an analysis of the concepts as published by various authors (it was conceived as highest value, average value, type, the most common phenomena, etc.). He pointed out, moreover, the difference between the norm concepts in analytical biology, and those in synthetic biology. In the last-mentioned case a reconsideration of the type-concept will be indispensable.

Meyer-Abich discussed the type-concept in several of his publications (cf.

Meyer-Abich, 1934, 1943, 1950, 1963, 1967). He compared it with a Platonic Idea, and arrived at the conclusion that type and natural system are Platonisms in modern biology. His view, and that of Troll, are probably the cause of much misunderstanding concerning typology among modern taxonomists. Meyer-Abich defined a type as the common characters of a taxon (an abstraction from nature), and an ideal schematic construction. Evidently, his Idea-concept is Kantian and not Platonic (cf. Kant, Critik der Reinen Vernunft, Zweite Abteilung, Erstes Buch, Erster Abschnitt: Von den Ideen überhaupt). As mentioned above, the type-concept in eighteenth century biology was based on Leibniz's substantial form, which concept, in its turn, was influenced by Aristotle and Scholasticism. Meyer-Abich distinguished between static and dynamic types. Phyla are called archetypes by him. He formulated a fundamental typological law, according to which, in a natural group, no organisms are found presenting all characters exclusively in the primitive state, nor are organisms found presenting all characters exclusively in the differentiated state. (Among all the possible combinations of characters only those are realized which fit into the environmental types). Meyer-Abich discussed also the problem of the origins of new types (in this discussion, his types are apparently not identical with Ideas).

Koort (1936, 1938) published a detailed contribution to the logic of the type-concept, in which he studied the type-concept in Goethe's works on natural history, the type-concept in traditional logic, and the type-concept in the humanities. He pointed out the synthetic character of the type-concept in these cases, and discussed the ideal type and the type as concept of the essence (as used in the humanities).

Brock (1939) made an attempt to develop the methodological foundations of an idealistic biology, on the analogy of idealistic morphology. He introduced the animal/environment/monad-concept and the monadology-concept, on the analogy of the type- and homology-concepts.

Danser (1940, 1950) studied the principles of typological and phylogenetic systematics. According to him a type or ground-plan is an imaginary living being which unites, in itself, all primitive character states (primitive without phylogenetic connotations). A ground-plan was not restricted by him to morphological characters, but could include, for instance, a general scheme of the life-cycle.

The English botanist Agnes Arber discussed Goethe's foliar theory of the appendicular organs of the stem, and Casimir De Candolle's partial-shoot theory (Arber, 1946, 1950). She further developed the last-mentioned theory, and introduced the hypothesis that the leaf is a partial-shoot which shows an urge towards whole-shoot characters. Although she was a vitalist (cf. Eyde, 1975), and expressly stated that her view had no phylogenetic implications, her hypothesis could be interpreted as referring to a type (of homonomous structures, a homotype) incorporating an evolutionary dimension.

Zangerl (1948) studied the methods of comparative morphology and the contribution it can make to the problems of evolution. He discussed the structural plan (plan of construction) and the morphotype (type), two methodological concepts which are closely related. He defined a morphotype as an ideally constructed form, the norm, within a group of organisms of essentially similar structural design, from which the actual forms of that group can be ideally derived. A morphological interpretation should precede a phylogenetic conclusion, and the concepts of both disciplines should not be interchanged.

Weber (1954: 205–207), in a discussion of the comparative morphology of insects, introduced the terms prototypal and metatypal for respectively primitive and derived character states (both states are not always clearly distinguished by him). A prototype unites in itself all prototypal character states of a taxon. Prototypes are at the same time palaeotypes which correspond with the ancestral forms in phylogeny. Eutypes present moreover the main derived character states of a taxon (Weber's definition is not unequivocal: eutypes are defined as presenting also the prototypal characteristics of a taxon).

Remane (1956) made a detailed study of type-concept and type-construction; his approach included morphological, taxonomic and phylogenetic aspects. He mentioned four methods of type construction, resulting in four different types: (1) A diagrammatic type, represented by a diagram, a scheme, a formula, etc.; it has an abstract character. (2) A generalized type which synthesizes the general characters of a taxon (all highly adaptive character states being reduced to a more general state). (3) A central type which synthesizes the central values of the various states of all characters. (4) A systematic type, in the construction of which much attention is paid to characters or character states of related groups, and groups of higher rank (of the same hierarchy); it synthesizes those character states which are essential from a taxonomic point of view. A systematic type can be used, in phylogeny, for the reconstruction of a common ancestor.

Sokal (1962) pointed to the importance of the type-concept in numerical taxonomy, and demonstrated the usefulness of empirical synthetic types, more or less corresponding with Smirnov's type and Remane's central type. He defined a type as a bounded multidimensional space contained within a multidimensional framework, the axis of which are the characters considered. This type allows for all variations of form within it (it is a space containing within it not only all the forms that have been studied to date, but also unoccupied regions into which additional forms could be placed at a later time; other additional forms might require the widening of the space).

Vogel (in: Jürgens & Vogel, 1965: 1–158) made a detailed comparative study of the type-concept in various disciplines. He distinguished three kinds of supraspecific types, viz., systematic type, life-form type, and functional type; and three kinds of infraspecific types, viz., constitutional type, social type, and racial type. He distinguished between type and type-model. A type-model is a type given concrete form; it can be constructed within a category, and depends on methodological principles (type-models can be average types, extreme types, static types, generalized types, diagrammatic types, ideal types, etc.). Besides total types, partial types can be distinguished (such as the type of a Tetrapod appendage). Vogel prepared the following general definition of a type: 'Ein morphologischer Typus ist ein allgemeines (nicht als einzelnes Individuum vorhandenes) Merkmalsganzes ('Integrat'), welches die grundlegenden und wesentlichen Charakteristika (auch, wenn einzelne typische Merkmale nicht allen, ja nicht einmal der Mehrzahl der Besonderungen zukommen) der durch ihn repräsentierten Gruppe von Individuen in einer allgemeinen Form besitzt, so dass es als die Formvielheit beherrschendes, einheitliches Gestaltungsprinzip (ohne jede metaphysische Bedeutung) alle Möglichkeiten der Ausformung seiner Besonderungen umfasst.'

Thompson (1966) demonstrated the current misconceptions with reference to typology. According to him, modern typology is evidently not based on Platonic views. He described the philosophical conception of the type in biology as the content of the taxonomic definition (this type not to be mistaken for the type-specimen).

In a short history of the biological model, Uschmann (1968) included Goethe's type-concept. He pointed out the fact that Goethe himself already characterized his 'Urpflanze' as a model. (In the same paper, Uschmann also classified Darwin's diagram of speciation by natural selection, in chapter 4 of The Origin of Species, and Haeckel's well-known phylogenetic trees with the biological models).

Schindewolf (1969) published a paper on the type-concept, with special reference to paleontology. He studied the relation of typology and population genetics, the type-concept in morphology and systematics, the type-concept in phylogenetics (including typogenesis), and accelerated type formation. He defined a type or plan of construction as those particular structures by which the largest natural groups of organisms are characterized. Schindewolf noted that this type-concept is completely different from a Platonic Idea, that it is an indispensable concept in systematics and comparative morphology, and that many investigators are unconsciously typologists, although they vigorously oppose typology.

Froebe (1971) discussed the methodological foundations of botanical typology. He defined a type as a model, and rejected the name 'idealistic morphology' because of its misleading character.

Voigt (1973) discussed the morphological type (the plan of construction) and the taxonomic type (Remane's systematic type), mainly in connection with the homology-concept. In his opinion, types are not necessarily restricted to morphology. He made an attempt to develop a general biological type-concept, which he provisionally defined as: a general theoretical model as an instrument of scientific thinking, obtained by abstraction of correlated characters, which are genetically coded and 'electively' realized in the phenotype.

Dullemeijer (1974) noted that the type-concept could easily be broadened by including the parameter time. In this way, a type would have static as well as dynamic aspects, and represent a modern scientific view.

It is interesting to add here that, in psychology, the type- as well as the archetype-concept are in use. In analytical psychology, the archetype-concept has, since 1919, been used to indicate an (especially innate) regulatory principle (a structural element of the unconscious) which, as archetypal image, can become conscious (cf. Jung, 1954). There is a distinct similarity between the archetypes from analytical psychology and the innate releasing mechanisms from ethology (cf. Heusser, 1976). In the study of psychological types, two different type-concepts are used, besides the archetype-concept, viz., (1) ideal, i.e. constructed types (cf. Jung, 1921), and (2) ordering categories which, in contradistinction to classes, are characterized by their imprecise, uncertain boundaries (cf. Hillman, 1980). The concept of the ideal type was studied by Rüstov (1953). It constitutes a group-concept, different from those obtained by abstraction of the common characters. It is obtained by heightening one or more views, and the combination of a great many characters in accordance with these views. Our concept of a typical Englishman, for instance, constitutes an ideal type. An ideal type is different from an average type and a normal type (the latter takes an intermediate position between ideal and average, although closer to the average type).

It may finally be remarked here that a logical typology was founded by Bertrand Russell (1872–1970) in 1908, in his essay on Mathematical Logic as Based on the Theory of Types (cf. Russell, 1956: 57–102). The subject was treated again by Whitehead & Russell (1910, introduction: Chapter 2). Russell defined a type as 'the range of significance of a propositional function, i.e., as the collection of arguments for which the said function has values.' His type-concept was applied to psychological typology by Hempel & Oppenheim (1936).

2.8. *Summary of the history of the type-concept*

In view of the development, in the next chapter, of a type as model of evolution and standard of higher classification, the following summary of the preceding subsections has been prepared.

The type-concept was introduced in biology in the course of the eighteenth century. In the history of ideas, it can be traced back to the idea of (substantial) form in Leibniz and Aristotle. The type-concept has been used both in comparative anatomy, and in the study of the homonomous structures of one individual. In the course of the nineteenth century, the type-concept developed in the direction of a mental construction, a model for the use of comparative morphology, that can be composed of homonomous elements. The use of the type-concept in systematic zoology (type-species, type-specimen) can be traced back to the use in morphology. In the twentieth century, a type has generally been understood as a model, a standard for the study of systematics, morphology, ecology and biology (it has also been used as an ordering category). Sever-

132

al methods have been developed for the construction of types, in which either primitive, typical, normal or average character states are used. A type in the modern sense can include ontogenetic as well as phylogenetic time. Some biologists have wrongly believed that types are identical with Platonic Ideas; they preferred an intuitive understanding of the type. In one case (Teichmüller, 1877), the type was developed as the concept of an innate genotypic pattern with evolutionary potentialities. In this sense, the concept approximates its original meaning and meets at the same time the requirements of a modern evolutionary approach.

3. The type as model of evolution and standard of higher classification

As mentioned in the preceding sections, the aim of the present study is to lay the theoretical foundations for the construction of a model of the evolutionary potentialities of a taxon (i.e. the innate genotypic pattern underlying evolution, and its manifestation in space and time), a model which can serve as a norm or standard for higher classification. Evidently, such a standard should be based on a complete comparative morphology (taking into account the data from other biological disciplines), and not on an arbitrary selection of characters.

A model in this sense constitutes a further development of the type-concept in morphology. In common with Richard Owen (1847, 1848), it is called archetype, on the one hand to preclude ambiguity (because of the use of a different type-concept in systematics), on the other hand because an archetype was characterized of old by immanent potentialities manifesting and developing themselves in space and time. This archetype should be represented by a long series of illustrations (mainly diagrams), accompanied by detailed explanations. It should include: a survey of characters and character states; the ontogenetic and phylogenetic aspects of the changes of character states; an analysis of radical changes viewed in connection with the adaptation to new ways of life (such as the adaptation to terrestrial life); and hypothetic models of evolutionary mechanisms. It could also include the relation between form and function, and the correlation of the elements of a functional component.

For the construction of a dynamic type, the recognition of homologies, and the distinction of character states are prerequisite. A first distinction between primitive (ancestral) and derived character states is typological, and one should admit that this discrimination remains typological unless a model is prepared of the evolutionary mechanisms underlying these changes of character states. Most distinctions, made by modern phylogenetic systematics, between plesiomorphous and apomorphous character states, are in fact typological distinctions between primitive and derived character states.

Another prerequisite for the construction of an archetype is constituted by the recognition of homonomous elements in a structural plan. As mentioned

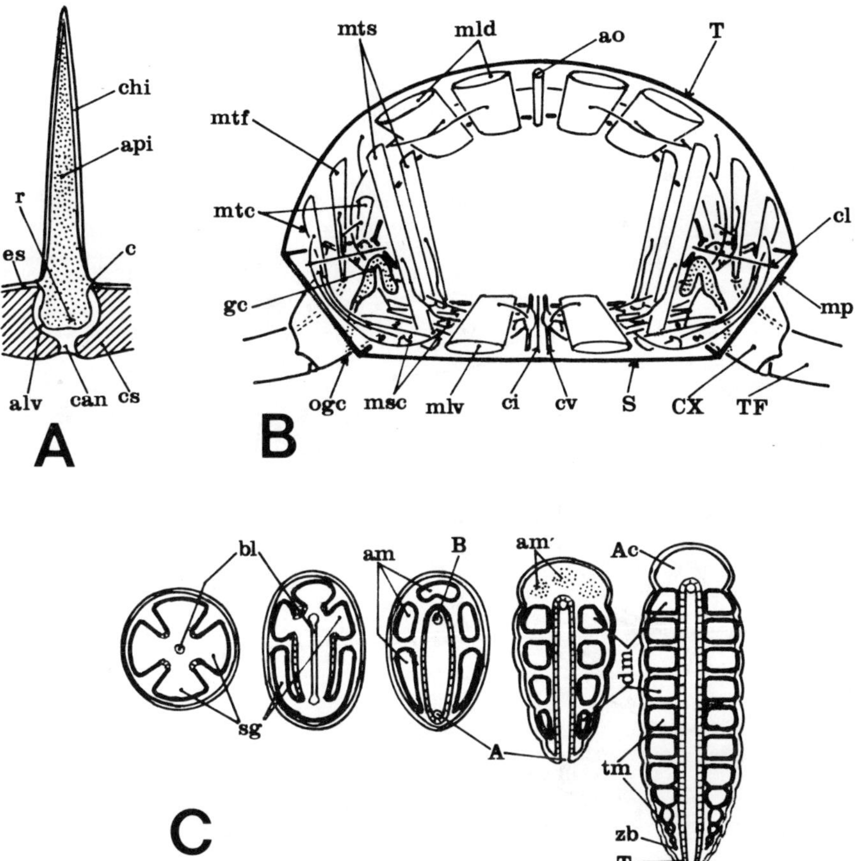

Fig. 5.4. Schematic representations which can form part of an archetype. A: homotype of an actinopilinous seta of Actinotrichid mites (redrawn after Grandjean, 1947); *alv*: alveolus; *api*: actinopilinous core; *c*: collar; *can*: canal; *chi*: chitinous part of seta; *cs*: chitonostracum (procuticle); *es*: epiostracum (epicuticle); *r*: root. B: homotype (merotype) of an arthropod metamere (redrawn after Chaudonneret, 1964); *ao*: aorta; *ci*: unpaired nerve cord; *cl*: lateral nerve cord; *cv*: paired ventral nerve cord; *CX*: coxa; *gc*: coxal gland (nephridium); *mld*: dorsal longitudinal muscles; *mlv*: ventral longitudinal muscles; *mp*: pleural membrane; *msc*: sterno-coxal muscles; *mtc*: tergo-coxal muscles; *mtf*: levator muscle of trochanterofemur; *mts*: tergo-sternal muscles; *ogc*: orifice of coxal gland; *S*: sternite; *T*: tergite; *TF*: trochanterofemur. C: origin of metamerism according to Remane (redrawn after Chaudonneret, 1964); *A*: anus; *Ac*: acron; *am*: archimetameres; *am '*: regressive archimetameres; *B*: mouth; *bl*: blastopore; *dm*: deutometameres; *sg*: gastric sacs; *T*: telson; *tm*: tritometameres; *zb*: proliferation zone.

above, Owen (1848) introduced the term homotype for the type of homonomous structures. An example of a homotype is constituted by the schematic figure of a seta of an Actinotrichid mite, reproduced in Fig. 5.4 A. A special kind of homotype is constituted by the type of a segment, which could be called

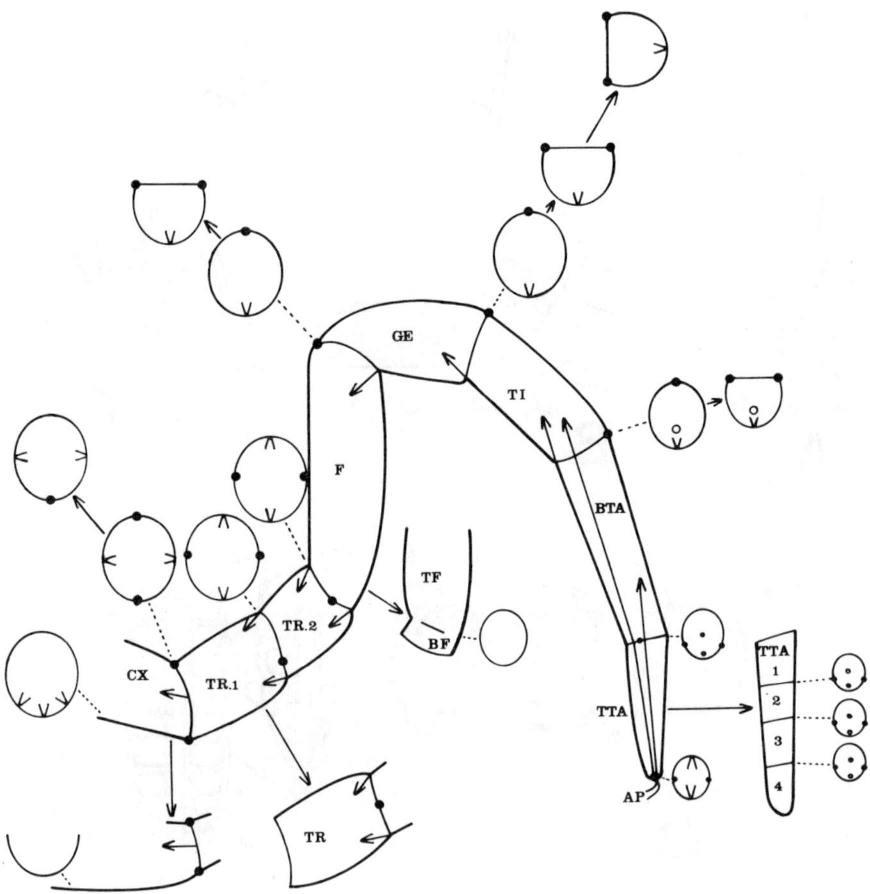

Fig. 5.5. Model of the evolution of leg III in Cryptognomae (Chelicerata). The ancestral leg III is supposed to be composed of: free coxa, trochanter 1, trochanter 2, femur, genu (patella), tibia, basitarsus, telotarsus and apotele. In the figure of the entire leg, articulation-points are represented by dots, tendons by arrows. The model demonstrates the evolution of a fixed coxa, the suppression of trochanter 2, the evolution of a basifemur, and the evolution of four telotarsi. The model demonstrates also the evolution of the separate joints. The schematic sections of the ancestral joints are connected, by dotted lines, with the joints in question (in the figure of the entire leg). The direction of the evolution is indicated by an arrow. In the schematic sections an articulation-point is represented by a dot, a tendon by a v-shaped symbol. The sections demonstrate the evolution of a rocking joint between coxa and trochanter, the evolution of hinge joints with flattened dorsal surface and two main articulation-points, and the evolution of torsion in the genu/tibia articulation. (After Van der Hammen, 1979). (The direction of the coxal evolution could be the reverse).

merotype (Fig. 5.4 B). In Fig. 5.4 A, B, the schematic figures are still static, because the evolutionary aspects are not included. A dynamic archetype should, for instance, include hypotheses with reference to the origin of metamerism (Fig. 5.4 C) and the subsequent differentiation of segments.

The dynamic aspects of an archetype are most clearly represented by ontophylogenetic diagrams representing the evolution of one character (or a character complex), in which phylogenetic and ontogenetic time are on the horizontal and the vertical axis respectively (cf. Van der Hammen, 1981a; Chapter V, Figs. 4–8). Ontophylogenetic diagrams can elucidate details of a more complex evolutionary scheme. Fig. 5.5, for instance, represents a model of the evolution of leg III in Cryptognomae (a chelicerate subclass). The evolution of trochanter 2 of this leg in Anactinotrichid mites (the largest group of Cryptognomae) is illustrated, in a more detailed way, in the ontophylogenetic diagram of Fig. 4.7.

An essential part of an archetype is constituted by hypothetical models of the evolutionary mechanisms underlying the changes in character states. The current evolutionary model is neo-Darwinian, i.e. based on the co-operation of genetic variability (mainly random mutations) and natural selection. In my paper on Numerical Changes and Evolution (Van der Hammen, 1981), I have demonstrated that numerical evolutionary changes (regressions as well as multiplications) can be the result of various types of changes in gene regulation (changes in systems of interactions); there appears, moreover, to be a genetically coded order in the manifestation of these evolutionary changes, whilst the changes themselves can be co-ordinated with changes in the environment. As mentioned above, the essence of this evolutionary model was predicted, as early as 1877, by Teichmüller. There are some indications (cf. Van der Hammen, 1981a: 26–28, Fig. 10) that a period of change of fundamental numbers, was preceded by a period in which the fundamental numbers arose from more chaotic numbers (apparently by the development of individually coded elements from elements coded collectively). This stage must have been preceded by an important first stage in the evolution of phyla: the origination of organisms composed of homonomous elements. As mentioned in section 2 of the present paper, an essential part of typological morphology was devoted to the comparative study of similar elements in an individual, and the recognition of homotypes. This most important aspect of morphology, which must be closely connected with the origin of new types or phyla, has been neglected by evolutionary biologists. The origin of segmentation, the further development of homonomous segmentation by anamorphosis, the origin of epimorphosis, and the development of heteronomous from homonomous segments, constitute phenomena which must be attributed to the early stages of evolution. Similar stages can, for instance, be distinguished in the numerical evolution of chaetotaxy (Fig. 2.12). Evidently, the various stages of numerical evolutions of various kinds of elements do not necessarily coincide. In Chelicerata, the first stabilization of the numbers of body-segments (the development of the so-called prototactic numbers) must have been attained at the end of the Cambrian, whilst primitive chaotic numbers of setae are supposed to be still present in some extant primitive species.

An important aspect of the archetype is constituted by its 'stratified' condition, i.e. the possibility to distinguish different layers or different adaptive periods or stages in the history of a group. In Chelicerata, an important transition is, for instance, constituted by the origin of terrestrial life (in most groups this transition took place before the Carboniferous). In aquatic Chelicerata, respiration must have been cutaneous, or have taken place by means of gills; in terrestrial Chelicerata, it is either cutaneous or takes place by tracheae and/or book-lungs. Fertilization must have been external in aquatic Chelicerata, internal in terrestrial Chelicerata. Digestion was certainly internal in aquatic Chelicerata, whilst external digestion could evolve only in the terrestrial period. The initial period of terrestrial life was probably restricted to the soil: sperm was deposited on the substratum in the shape of spermatophores. The development of sperm transfer by gonopods, or through copulation, opened up the possibility to colonize different niches. The same applies to the development of web-construction in Spiders. The distinction, in the history of a group, of a primitive type (plesiotype) and various superimposed adaptive types (telotypes), enables us to attribute a certain evolutionary change to a certain period of time, and to recognize a certain hierarchy of changes. It may be remarked here that, in this connection, the relation between form and function, the correlation of the elements of a functional component, and a construction-morphological view are of considerable importance.

A comprehension of the evolution of a natural group, and the interpretation of its character states, requires also an insight in the types that have not been realized (i.e. the gaps or discontinuities between groups) and the types that are no more extant (i.e. the extinct groups). I confine myself to a few remarks. Much has been published on the extinction of species and groups (cf. Hoernes, 1911; Remane, 1956: 308–309). The construction of extincto-types (based on the common characters of extinct groups) could indeed be interesting, just as the construction of hiatotypes (representing the important gaps between the different groups).[6] The last-mentioned types are also important because classification is partly based on discontinuities.

The construction of an archetype, according to the methods and principles demonstrated above, demands a considerably factual knowledge obtained by detailed comparative studies and complete descriptions. An extensive study of this kind prevents, more or less, the results of an investigation from being subjectively influenced by the restrictions of the observation program. The archetype can constitute an important standard in the pursuit of an objective classification, and a norm by which the hierarchical value of characters and character states can be tested. A complete archetype does indeed constitute the standard by which one can test whether a classification can be rightly called phylogenetic (many classifications now introduced as phylogenetic are based on an arbitrary

---

[6] The fascinating problem of morphologies that do not occur in nature was investigated by Gould (1984). He studied morphological channeling by structural constraint in shells of the Gastropod genus *Cerion*.

selection of characters, and an arbitrary interpretation of character states; they can be more artificial than a classification introduced as typological). All classifications based on the same archetype, whether typological, numerical or phylogenetic, must indeed be more or less similar, although not necessarily identical.

The archetype as model of the evolutionary mechanisms should contain all information available in this field. According to the present state of our knowledge, this information will be fragmentary. The model will demonstrate that most classifications are indeed typological, or are at most based on a phylogenetic arrangement of typological interpretations. This is no disqualification as long as the archetype is constructed as detailed as possible, and consequently constitutes the highest possible standard.

References

Anonymous, 1843. [Reports of researches undertaken at the request of the Association and published in its Transactions]. – Rep. Twelfth Meeting Brit. Ass. Adv. Sci. 1842: xvi–xix.

Anonymous, 1843a. Recommendations of researches in science involving grants of money, adopted by the general committee at the Twelfth Meeting. – Rep. Twelfth Meeting Brit. Ass. Adv. Sci. 1842: xx–xxiv.

Anonymous, 1843b. Report of a committee appointed 'to consider the rules by which the nomenclature of zoology may be established on a uniform and permanent basis'. – Rep. Twelfth Meeting Brit. Ass. Adv. Sci. 1842: 105–121.

Anonymous, 1843c. Series of propositions for rendering the nomenclature of zoology uniform and permanent, being the report of a committee for the consideration of the subject appointed by the British Association for the Advancement of Science. – Ann. Mag. Nat. Hist. 11: 259–275.

Anonymous, 1892. The Code of Nomenclature adopted by the American Ornithologists' union. – New York: iv + 72 p.

Arber, A., 1946. Goethe's botany. The Metamorphosis of Plants (1790) and Tobler's Ode to Nature (1782). – Chronica Botanica 10: 64–126.

Arber, A., 1950. The natural philosophy of plant form. – Cambridge: xiv + 248 p. (Reprint: Darien, Conn., 1970).

Aristotle: The metaphysics, books I–IX. With an English translation by Hugh Tredennick. – Cambridge, Mass., London (The Loeb Classical Library 271), 1980: xi + 473 p.

Baer, K.E. von, 1827. Beiträge zur Kenntnis der niedern Thiere. – Nova Acta Phys.-Med. Acad. Caesar. Leop.-Carol. Natur. Curios. 13(2): 523–762, 881–882.

Baer, K.E. von, 1828. Über Entwicklungsgeschichte der Thiere. Beobachtung und Reflexion. Erster Theil. – Königsberg: xxiv + 272 p.

Bliedner, A., 1901. Goethe und die Urpflanze. – Frankfurt: iv + 76 p.

Blumenthal, A. von, 1928. Typos und Paradeigma. – Hermes, Zeitschrift für Klassische Philologie 63: 391–414.

Borman, C.V., W. Franzen, A. Krapiec & L. Oeing-Hanhoff, 1972. Form und Materie (Stoff). – In: J. Ritter (ed.), Historisches Wörterbuch der Philosophie 2: 977–1030.

Bräuning-Oktavio, H., 1956. Vom Zwischenkieferknochen zur Idee des Typus. Goethe als Naturforscher in den Jahren 1780–1786. – Nov. Act. Leopold. (N.F.) 18(126): 1–144.

Broad, C.D., 1979. Leibniz. An introduction (edited by C. Lewy). – Cambridge (second edition): xii + 175 p.

Brock, F., 1939. Typenlehre und Umweltforschung. Grundlegung einer idealistischen Biologie. – Bios 9: viii + 40 p.

Bruno, G.: Von der Ursache, dem Prinzip und dem Einen (transl. A. Lasson, intr. W. Beierwaltes, ed. P.R. Blum). – Hamburg (Philosophische Bibliothek, 21), 1977 (5th edition): lxiv + 20 + 152 p.

Buffon, G.-L. Leclerc de, 1749. Histoire naturelle, générale et particulière, avec la description du Cabinet du Roi. Tome second. – Paris: iv + 603 p.

Buffon, G.-L. Leclerc de & L.J.M. Daubenton, 1753. Histoire naturelle, générale et particulière, avec la description du Cabinet du Roi. Tome quatrième. – Paris: xvi + 544 p.

Carus, C.G., 1861. Natur und Idee oder das Werdende und sein Gesetz. Eine philosophische Grundlage für die specielle Naturwissenschaft. – Wien: x + 492 p.

Chaudonneret, J., 1964. Une unité structurale: le métamère. – Cahiers d'Études Biologiques 13–15: 77–101.

Cuvier, G., 1799. Leçons d'anatomie comparée 1. Contenant les organes du mouvement. – Paris: xxxii + 522 p.

Cuvier, G., 1817. Le règne animal distribué d'après son organisation, pour servir de base à l'histoire naturelle des animaux et d'introduction à l'anatomie comparée 1. – Paris: xxxviii + 540 p.

Cuvier, G., 1829. Le règne animal distribué d'après son organisation, pour servir de base à l'histoire naturelle des animaux et d'introduction à l'anatomie comparée 1. – Paris (second edition): xxxviii + 584 p.

Cuvier, G., 1836–1849. Le règne animal distribué d'après son organisation, pour servir de base à l'histoire naturelle des animaux, et d'introduction à l'anatomie comparée. Édition accompagnée de planches gravées, représentant les types de tous les genres, les caractères distinctifs des divers groupes et les modifications de structure sur lesquelles repose cette classification 1–17. – Paris (third edition).

Danser, B.H., 1940. Typologische en phylogenetische systematiek. – Vakblad voor Biologen, 21: 137–145.

Danser, B.H., 1950. A theory of systematics. – Bibliotheca Biotheoretica 4(3): 113–180.

Darwin, C., 1872. The origin of species by means of natural selection, or the preservation of favoured races in the struggle for life. – London (sixth edition): xxii + 458 p.

Descartes, R.: Œuvres de Descartes publiées par Charles Adam & Paul Tannery. Vol. 7: Meditationes de prima philosophia. Vol.: 9(1): Méditations. Traduction française. – Paris (1964).

Diderot, D., 1754. Pensées sur l'interprétation de la nature. – Paris: ii + 218 p. [Published anonymously; place of publication not mentioned on the title-page. I have used also the following edition, which has an introduction and additional notes: Œuvres complètes. Edition chronologique. Le Club Français du Livre, vol. 2 (1969): 707–774].

Dullemeijer, P., 1974. Concepts and approaches in animal morphology. – Assen (Van Gorcum & Comp. B.V.): xii + 264 p.

Eyde, R.H., 1975. The foliar theory of the flower. – American Scientist 63: 430–437.

Franzen, W., K. Georgulis & H.M. Nobis, 1972. Entelechie. – In: J. Ritter (ed.), Historisches Wörterbuch der Philosophie 2: 506–509.

Froebe, H.A., 1971. Die wissenschaftstheoretische Stellung der Typologie. – Ber. Deutsch. Bot. Ges. 84: 119–129.

Geoffroy Saint-Hilaire, E., 1818. Philosophie anatomique des organes respiratoires sous le rapport de la détermination et de l'identité de leurs pièces osseuses. – Paris: 1 + 518 p.

Geoffroy Saint-Hilaire, E., 1830. Principes de philosophie zoologique, discutés en mars 1830, au sein de l'Académie Royale des Sciences. – Paris: 227 p.

Glanville, J.: Scepsis scientifica: or, confest ignorance, the way to science; in an essay of the vanity of dogmatizing, and confident opinion, ed. by J. Owen. – London, 1885: lxx + 218 p.

Goethe, J.W. von: Gedenkausgabe der Werke, Briefe und Gespräche. Herausgegeben von Ernst Beutler. – Zürich, Stuttgart (Artemis Verlag). Vol. 11. Italienische Reise. Annalen. Intr., ed. E. Beutler, 1962 (second edition): 1181 p. Vol. 18. Briefe der Jahre 1764–1786. Intr. E. Beutler, ed. E. Damin, 1965 (second edition): 1295 p. Vol. 19. Briefe der Jahre 1786–1814. Intr., ed. H. Ostertag, 1962 (second edition): 639 p. Suppl. vol. 2. Tagebücher. Ed. P. Boerner, 1964: 955 p.

Goethe, J.W. von: Die Schriften zur Naturwissenschaft. Vollständige mit Erläuterungen versehene Ausgabe herausgegeben im Auftrage der Deutschen Akademie der Naturforscher (Leopoldina)

zu Halle von Rupprecht Matthaei, Wilhelm Troll, Lothar Wolf, Dorothea Kuhn & Wolf von Engelhardt. – Weimar (Hermann Böhlaus Nachfolger). Erste Abteilung: Texte. Vol. 9. Morphologische Hefte. Bearbeitet von Dorothea Kuhn. – Weimar 1954: xii + 389 p., 32 pls. Vol. 10. Aufsätze, Fragmente, Studien zur Morphologie. Bearbeitet von Dorothea Kuhn. – Weimar 1964: viii + 408 p., 27 pls.
Zweite Abteilung: Ergänzungen und Erläuterungen, Vol. 9A. Zur Morphologie. Von den Anfängen bis 1795. Ergänzungen und Erläuterungen. Bearbeitet von Dorothea Kuhn. – Weimar 1977: xxviii + 607 p., 15 pls. Vol. 9B. Zur Morphologie. Von 1796 bis 1815. Ergänzungen und Erläuterungen. Bearbeitet von Dorothea Kuhn. – Weimar 1986: xxx + 602 p., 8 pls.

Goethe, J.W. von: Corpus der Goethezeichnungen, vol. 5B. Die Naturwissenschaftlichen Zeichnungen mit Ausnahme der Farbenlehre. Ed. D. Kuhn, O. Wagenbreth & K. Schneider-Carius. – Leipzig (E.A. Seemann Buch- und Kunstverlag) 1967: 132 p. 164 figs.

Gould, S.J., 1984. Morphological channeling by structural constraint: convergence in styles of dwarfing and gigantism in Cerion, with a description of two new fossil species and a report on the discovery of the largest Cerion. – Paleobiology 10: 172–194, Figs. 1–14, Tab. 1–8.

Grandjean, F., 1947. L'origine de la pince mandibulaire chez les Acariens actinochitineux. – Arch. Sci. Phys. Natur. Genève (5) 29: 305–355.

Haecker, V., 1927. Goethes morphologische Arbeiten und die neuere Forschung. – Jena: vi + 98 p.

Hammen, L. van der, 1977. A new classification of Chelicerata. – Zool. Meded. Leiden 51: 307–319, Fig. 1, Tab. 1–3.

Hammen, L. van der, 1979. Comparative studies in Chelicerata I. The Cryptognomae (Ricinulei, Architarbi and Anactinotrichida). – Zool. Verh. Leiden 174: 1–62, Figs. 1–31.

Hammen, L. van der, 1981. Type-concept, higher classification and evolution. – Acta Biotheoretica 30: 3–48, Figs. 1–7.

Hammen, L. van der, 1981a. Numerical changes and evolution in Actinotrichid mites (Chelicerata). – Zool. Verh. Leiden 182: 1–48, Figs. 1–13. [Cf. essay II].

Hammen, L. van der, 1982. Comparative studies in Chelicerata II. Epimerata (Palpigradi and Actinotrichida). – Zool. Verh. Leiden 196: 1–70, Figs. 1–31.

Hammen, L. van der, 1983. Unfoldment and manifestation. The natural philosophy of evolution. – Acta Biotheoretica 32: 179–193. [Cf. essay I].

Hammen, L. van der, 1985. Comparative studies in Chelicerata III. Opilionida. – Zool. Verh. Leiden 220: 1–60, Figs. 1–34.

Hammen, L. van der, 1986. On some aspects of parallel evolution in Chelicerata. – Acta Biotheoretica, 35: 15–37, Figs. 1–9. [Cf. essay IV].

Hammen, L. van der, 1986a. Comparative studies in Chelicerata IV. Apatellata, Arachnida, Scorpionida, Xiphosura. – Zool. Verh. Leiden 226: 1–52, Figs. 1–23.

Hammen, L. van der, 1986b. Acarological and arachnological notes. – Zool. Meded. Leiden 60: 217–230, Figs. 1–3.

Hansen, A., 1919. Goethes Morphologie (Metamorphose der Pflanzen und Osteologie). Ein Beitrag zum sachlichen und philosophischen Verständnis und zur Kritik der morphologischen Begriffsbildung. – Giessen: iv + 200 p.

Harris, C.L., 1981. Evolution, genesis and revelations. With readings from Empedocles to Wilson. – Albany (State University of New York Press): xii + 339 p., Figs. 1–39, Tab. 1–8.

Hempel, C.G. & P. Oppenheim, 1936. Der Typusbegriff in Lichte der neuen Logik. Wissenschaftstheoretische Untersuchungen zur Konstitutionsforschung und Psychologie. – Leiden: viii + 130 p.

Herder, J.G. von, 1784. Ideen zur Philosophie der Geschichte der Menschheit 1. – Leipzig. (I have used the fourth edition, edited by H. Luder, 1841).

Heusser, H. (ed.), 1976. Instinkte und Archetypen im Verhalten der Tiere und im Erleben des Menschen. – Wege der Forschung, Darmstadt 80: vi + 402 p.

Heyden, C. von, 1826. Versuch einer systematischen Eintheilung der Acariden. – Isis (Oken) 1826: 608–613.

Hillman, J., 1980. Egalitarian typologies versus the perception of the unique. – In: A. Portmann & R. Ritsema (ed.), Oneness and variety (Eranos Yearbook 45): 221–279.

Hoernes, R., 1911. Das Aussterben der Arten und Gattungen sowie der grösseren Gruppen des Tier- und Pflanzenreiches. – Festschr. k.k. Karl-Franzens-Univ. Graz 1910/11: viii + 255 p.

Jürgens, H.W. & C. Vogel, 1965. Beiträge zur menschlichen Typenkunde. – Stuttgart: viii + 255 p.

Jung, C.G., 1921. Psychologische Typen. – Zürich: 708 p. [= Gesammelte Werke 6 (1960)].

Jung, C.G., 1954. Von den Wurzeln des Bewusstseins. – Zürich 681 p. [Cf. Gesammelte Werke 8 (1967): 185–267; 9(I) (1976): 11–51, 67–123].

Kant, I., 1787. Critik der reinen Vernunft. – Riga (second edition). [The following edition was used by me: I. Kant, Werke in zehn Bänden (ed. W. Weischedel), vol. 3–4: Kritik der reinen Vernunft. – Darmstadt, 1975 (fourth edition)].

Kant, I., 1790. Critik der Urtheilskraft. – Berlin, Libau. [The following edition was used by me: I. Kant, Werke in zehn Bänden (ed. W. Weischedel), vol. 8: Kritik der Urteilskraft und Schriften zur Naturphilosophie. – Darmstadt, 1975 (fourth edition)].

Klaauw, C.J. van der, 1931. Normaal, norm en normbegrip in de biologie. – Vakblad voor Biologen 12: 173–183.

Koort, A., 1936. Beiträge zur Logik des Typusbegriffs. Teil I. – Acta et Commentationes Universitatis Tartuensis (Dorpatensis) (B) 38(4): 1–138.

Koort, A., 1938. Beiträge zur Logik des Typusbegriffs. Teil II. – Acta et Commentationes Universitatis Tartuensis (Dorpatensis) (B) 39(1): i–iv, 139–263.

Kuhn, D., 1967. Empirische und ideelle Wirklichkeit. Studien über Goethes Kritik des französischen Akademiestreites. – Graz, Wien, Köln (Neue Hefte zur Morphologie 5): 319 p.

Latreille, P.A., 1802. Histoire naturelle, générale et particulière des Crustacés et des Insectes 3. – Paris: xii + 468 p.

Latreille, P.A., 1804. Histoire naturelle, générale et particulière des Crustacés et des Insectes 7. – Paris: 415 p.

Latreille, P.A., 1810. Considérations générales sur l'ordre naturel des animaux composant les classes des Crustacés, des Arachnides, et des Insectes; avec un tableau méthodique de leur genres, disposés en familles. – Paris, 444 p.

Leibniz, G.W.: Die philosophischen Schriften (ed. C.I. Gerhardt) 2. – Berlin, 1879: viii + 594 p.

Leibniz, G.W.: Lettres de Leibniz à Arnauld d'après un manuscrit inédit (ed. G. Lewis). – Paris, 1952: iv + 116 p.

Locke, J., 1690. An essay concerning humane understanding. In four books. – London [I have used the edition of 1894 (ed. A.C. Fraser), repr. New York, 1959, 2 vol.].

Lovejoy, A.O., 1936. The great Chain of Being. A study of the history of an idea. The William James lectures delivered at Harvard University, 1933. – Cambridge, Mass. [I have used the fourteenth printing: Cambridge, Mass., 1978: xii + 382 p.].

Maclise, J., 1847. Comparative osteology: being morphological studies to demonstrate the archetype skeleton of vertebrated animals. – London: 180 p.

Mayr, E., 1969. Principles of systematic zoology. – New York: xiv + 428 p.

Mayr, E., E.G. Linsley & R.L. Usinger, 1953. Methods and principles of systematic zoology. – New York, Toronto, London: x + 336 p.

Meinhardt, H., 1972. Idee, Antike. – In: J. Ritter (ed.), Historisches Wörterbuch der Philosophie 2: 55–66.

Meyer-Abich, A., 1934. Ideen und Ideale der biologischen Erkenntnis. Beiträge zur Theorie und Geschichte der biologischen Ideologien. – Bios, Leipzig 1: 1–202.

Meyer-Abich, A., 1943. Beiträge zur Theorie der Evolution der Organismen I. Das typologische Grundgesetz und seine Folgerungen für Phylogenie und Entwicklungsphysiologie. – Acta Biotheoretica 7: 1–80.

Meyer-Abich, A., 1950. Beiträge zur Theorie der Evolution der Organismen II. Typensynthese durch Holobiose. – Bibliotheca Biotheoretica 5: 1–206.

Meyer-Abich, A., 1963. Geistesgeschichtliche Grundlagen der Biologie. – Stuttgart: viii + 322 p.

Meyer-Abich, A., 1967. Der Typus-Begriff in der Biologie. – Leopoldina (3) 12: 179–200.

Michéa, R., 1958. Goethe et les évolutionnistes français du XVIIIe siècle. – In: Goethe et l'Esprit Français (Publ. Fac. Lettr. Univ. Strasbourg 137: 129–149.

Micraelius, J., 1662. Lexicon philosophicum. Terminorum philosophis usitatorum. – Stettin (second edition). [Reprint: Düsseldorf, 1966].

Naef, A., 1919. Idealistische Morphologie und Phylogenetik. (Zur Methodik der systematischen Morphologie). – Jena: vi + 78 p.

Naef, A., 1931. Die Gestalt als Begriff und Idee. – In: L. Bolk, E. Kallius & W. Lubosch, Handbuch der Vergleichenden Anatomie der Wirbeltiere 1: 77–118.

Nicard, P., 1890. Étude sur la vie et les travaux de M. Ducrotay de Blainville. – Paris: 255 p.

Nisbet, H.B., 1967. Herder, Goethe, and the natural 'type'. – Publ. Engl. Goethe Soc. (N.S.) 37: 83–119.

Oken, L., 1807. Über die Bedeutung der Schädelknochen. Ein Programm beym Antritt der Professur an der Gesammt-Universität zu Jena. – Jena: 18 p.

Oken, L., 1847. Elements of physiophilosophy (transl. A. Tulk). – London (The Ray Society): xx + 665 p.

Owen, R., 1846. Lectures on the comparative anatomy and physiology of the vertebrate animals, delivered at the Royal College of Surgeons of England, in 1844 and 1846. Part I. Fishes. – London: xii + 308 p.

Owen, R., 1847. Report on the archetype and homologies of the vertebrate skeleton. – Rep. Sixteenth Meeting Brit. Ass. Adv. Sci. 1846: 169–340.

Owen, R., 1848. On the archetype and homologies of the vertebrate skeleton. – London: viii + 203 p.

Owen, R., 1858. Oken. – The Encyclopaedia Britannica or Dictionary of Arts, Sciences, and General Literature (eighth edition) 16 (Nav-Orn): 498–503.

Owen, R., 1894. The life of Richard Owen by his grandson The Rev. Richard Owen M.A. With the scientific portions revised by C. Davies Sherborn. Also an essay on Owen's position in anatomical science by the Right Hon. T.H. Huxley, F.R.S. 1, 2. – London: xvi + 409 p. (vol. 1), viii + 393 p. (vol. 2).

Peyer, B., 1950. Goethes Wirbeltheorie des Schädels. – Vierteljahrsschr. Naturf. Ges. Zürich 94 (Beih. 2/3): 132 p.

Philo Judaeus (= Alexandrinus): F.H. Colson & G.H. Whitaker (ed.), Philo. With an English translation 1–10. – London (The Loeb Classical Library).

Pico della Mirandola, G.: Johannes Picus, Opera omnia. – Basileae. [Reprint: Torino, 1971].

Plato: Platon, Œuvres complètes 1–14. – Paris (Collection des Universités de France).

Plotinus: Plotin, Ennéades 1–6. – Paris (Collection des Universités de France).

Remane, A., 1956. Die Grundlagen des natürlichen Systems, der vergleichenden Anatomie und der Phylogenetik. Theoretische Morphologie und Systematik I. – Leipzig (second edition): vi + 364 p. (Reprint: Koenigstein-Taunus, 1971).

Robinet, J.B., 1761–1766. De la nature 1–4. – Amsterdam; vol. 1(1761): xx + 456 p. (published anonymously); vol. 2(1763): xvi + 444 p.; vol. 3(1766): lvi + 288 p.; vol. 4(1766a): iv + 284 p.

Robinet, J.B., 1768. Vue philosophique de la gradation naturelle des formes de l'être, ou les essais de la nature qui apprend à faire l'homme. – Amsterdam: iv + 264 p.

Roger, J., 1965. Die Auffassung des Typus bei Buffon und Goethe. – Die Naturwissenschaften 52: 313–319.

Rouché, M., 1940. La philosophie de l'histoire de Herder. – Paris (Publ. Fac. Lettr. Univ. Strasbourg 93): viii + 720 p.

Rüstow, A., 1953. Der Idealtypus, oder die Gestalt als Norm. – Studium Generale 6: 54–59.

Russell, B., 1956. Logic and knowledge. Essays 1901–1950 (ed. R.C. Marsh). – London: xii + 382 p.

Schindewolf, O.H., 1969. Über den 'Typus' in morphologischer und phylogenetischer Biologie. – Abh. Akad. Wiss. u. Lit., Math.-Naturw. Klasse 1969: 55–131.

Schmid, G., 1930. Goethes Metamorphose der Pflanzen. – In: J. Walther (ed.), Goethe als Seher und Erforscher der Natur. Untersuchungen über Goethes Stellung zu den Problemen der Natur: 205–226, 313–319.

Schmid G., 1940. Goethe und die Naturwissenschaften. Eine Bibliographie. – Halle (Saale): xvi + 620 p.

Simpson, G.G., 1967, Principles of animal taxonomy. – New York, London (Columbia Biological Series 20): xiv + 247 p. (third printing). [The first edition was published in 1961].

Smirnov, E., 1925. The theory of type and the natural system. – Zeitschr. Indukt. Abstammungs- u. Vererbungslehre 37: 28–66.

Sokal, R.R., 1962. Typology and empiricism in taxonomy. – Journ. Theor. Biol. 3: 230–267.

Spinner, H., 1933. Goethes Typusbegriff. – Horgen-Zürich, Leipzig (Wege zur Dichtung 16): 273 p.

Spinoza, B. de: On the improvement of the understanding. The Ethics. Correspondence (transl., intr. R.H.M. Elwes). – New York, 1955 (reprint of the 1883 edition): xxii + 420 p.

Steiner, R., 1897. Goethes Weltanschauung. – Weimar. [I have used the fifth edition, Dornach, 1963: 223 p.].

Steiner, R., 1926. Goethes naturwissenschaftliche Schriften. – Dornach. [I have used the third edition, Dornach, 1973: 350 p. The separate chapters originally appeared, in the period 1883–1897, as introductions to Steiner's edition of Goethe's Naturwissenschaftliche Schriften].

Strickland, H.E., 1837. Rules for zoological nomenclature. – Mag. Nat. Hist. (N.S.) 1: 173–176.

Strickland, H.E., 1878. Rules for zoological nomenclature. – London: 28 p.

Teichmüller, G., 1877. Darwinismus und Philosophie. – Dorpat: viii + 90 p.

Thienemann, A., 1909. Die Stufenfolge der Dinge, der Versuch eines natürlichen Systems der Naturkörper aus dem achtzehnten Jahrhundert. Eine historische Skizze. – Zool. Annalen 3: 185–274.

Thomas, O., 1893. Suggestions for a more definite use of the word 'Type' and its compounds, as denoting specimens of a greater or less degree of authenticity. – Proc. Zool. Soc. Lond. 1893: 241–242.

Thompson, W.R., 1966. The status of species. – In: V.E. Smith (ed.), Philosophical Problems in Biology (St. John's University Philosophical Series 5): 67–126.

Troll, W., 1928. Organisation und Gestalt im Bereich der Blüte. – Berlin (Monographien aus dem Gesamtgebiet der Wissenschaftlichen Botanik 1): xvi + 413 p.

Troll, W., 1941. Gestalt und Urbild. Gesammelte Aufsätze zu Grundfragen der organischen Morphologie. – Leipzig (Die Gestalt, Abhandlungen zu einer allgemeinen Morphologie 2): viii + 181 p.

Turpin, P.J.F., 1820. Essai d'une iconographie élémentaire et philosophique des végétaux avec un texte explicatif. – In: J.L.M. Poiret, Flore médicale. Partie élémentaire 7bis: 200 p.

Turpin, P.J.F., 1837. Esquisse d'organographie végétale, fondée sur le principe d'unité de composition organique et d'évolution rayonnante ou centrifuge, pour servir à prouver l'identité des organes appendiculaires des végétaux et la métamorphose des plantes de Goethe. – In: Œuvres d'histoire naturelle de Goethe, comprenant divers mémoires d'anatomie comparée, de botanique et de géologie traduits et annotés par Ch.Fr. Martins, docteur en médecine, *Atlas*. – Paris, Genève: 5–70.

Uschmann, G. 1968. Die Naturgeschichte des biologischen Modells. – Nova Acta Leopoldina (N.F.) 33(184) (= J.-H. Scharf & G. Bruns (ed.), Biologische Modelle): 43–64.

Vicq d'Azyr, F., 1786. Traité d'anatomie et de physiologie avec des planches coloriées. Représentant au naturel les divers organes de l'homme et des animaux 1. – Paris: 123 + 111 p.

Vicq d'Azyr, F., 1805. Œuvres. Recueillies et publiées avec des notes et un discours sur sa vie et ses ouvrages, par Jacq. L. Moreau (de la Sarthe) 4. – Paris: 410 p.

Voigt, W., 1973. Homologie und Typus in der Biologie. Weltanschaulich-philosophische und erkenntnistheoretisch-methodologische Probleme. – Jena: 132 p.

Walsingham, Lord & J.H. Durant, 1896. Rules for regulating nomenclature with a view to secure a strict application of the law of priority in entomological work. – London: 18 p.

Weber, H., 1954. Grundriss der Insektenkunde. – Stuttgart (third edition): xi + 428 p.

Wells, G.A., 1978. Goethe and the development of science 1750–1900. – Alphen aan den Rijn (Science in History 5): xii + 161 p.

Whitehead, A.N. & B. Russell, 1910. Principia mathematica. – Cambridge. [I have used the edition of 1976, To *56, Cambridge: xlvi: 410 p.].

Wilkie, J.S., 1957. The idea of evolution in the writings of Buffon. – Annals of Science 12: 48–62, 212–227, 255–266.

Wilkie, J.S., 1959. Buffon, Lamarck and Darwin: The originality of Darwin's theory of evolution. – In: P.R. Bell (ed.), Darwin's Biological Work. Some Aspects Reconsidered: 262–307.

Wilkinson, E.M., 1953. Goethe's conception of form. – Proc. Brit. Acad. 37: 175–197.

Wolff, C.F., 1768. De formatione intestinorum. – Nov. Comm. Acad. Sci. Imp. Petropolitanae 12: 403–507.

Zangerl, R., 1948. The methods of comparative anatomy and its contribution to the study of evolution. – Evolution 2: 351–374.

Ziehen, T., 1930. Goethes naturphilosophische Anschauungen. – In: J. Walther (ed.), Goethe als Seher und Erforscher der Natur. Untersuchungen über Goethes Stellung zu den Problemen der Natur: 35–57, 303–306.

VI. A STRUCTURALIST APPROACH IN THE STUDY OF EVOLUTION AND CLASSIFICATION

1. Introduction

In essay V I have developed a type-concept as a model of the evolutionary potentialities of a taxon, taking into account ontogeny, phylogeny and evolutionary mechanisms. This type (or archetype) represents, in fact, a model of a system of transformation, i.e. a structure in the structuralist sense. Essay VI, devoted to structuralism, is a logical development of essay V, and includes also data from other essays (particularly essays II, III and IV). It is based on two papers, viz., a paper under the same title as the present essay (Van der Hammen, 1985b), and a paper entitled 'Structuralism in evolutionary biology and systematics'' (read at the Workshop Structuralism in Biology, Osaka, Japan, December 1986; cf. Van der Hammen, in press).

Structuralism (see Piaget, 1968; Lane, 1970) is essentially a successful method currently applied to research in a wide variety of disciplines, among which mathematics, physical sciences, anthropology, mythology, linguistics and literary criticism. It is rather a method (in the sense of procedure adopted for investigation) than a theory, although it depends ultimately also on certain theoretical assumptions. Because of the great diversity of disciplines to which the structuralist method is applied, structuralism is not easy to define; in all disciplines, it employs the notion 'structure' (in the sense of set of elements between which relations are defined), and its intention is to go beyond the description of the observational data, in the direction of the laws underlying the phenomena in which it is concerned. Relations at the level of observable reality are regarded as reflections of relations at the level of a structure below or behind reality; it is this 'deep' structure (acting as a structuring force) which structuralism attempts to discover.

According to Piaget (1968) the notion of structure is comprised of three key ideas: the idea of wholeness, the idea of transformation, and the idea of self-regulation. The most distinct feature of structuralism is that it gives emphasis not to aggregates (or composites of independent elements) but to wholes. Structuralism attempts to study the relations between the elements, and it is in terms of these that the structure is defined. The relations between the elements are law-like regularities (the so-called laws of composition) and these include the so-called laws of transformation, by which one particular structured con-

146

figuration changes into another. All structures are, in fact, systems of transformation (which makes the question of origin inevitable). Systems of transformation of this kind depend upon mechanisms of self-regulation (entailing self-maintenance and a certain closure), and the basic mechanisms of self-regulation include regulations in the cybernetic sense of the word. (In biology, self-regulation includes reproduction).

Structuralism does not look for causal explanations (these belong to the domain of different methods); one of its main purposes is to prepare a complete inventory of structural relations (among which binary oppositions). In addition to that, structuralism is mainly concerned with structural relations existing at the same time (because of the idea of wholeness); history is seen as a mode of development of a particular system, whose present nature must be fully known before any account can be given of its evolution. It may finally be remarked here that, in many cases, a hierarchy of structures can be imagined participating in the process of generation of a surface structure.

Webster & Goodwin (1981, 1982) criticized the current neo-Darwinian view of biology, in which the organism as a real entity has virtually no place. They argued that a structuralist approach in the study of the organismic domain might favourably transform modern biology. They found the germs of this approach in the pre-Darwinian school of comparative anatomy and embryology (from Cuvier to Owen) and its concern with laws of form, as well as in some fringe figures outside the mainstream of twentieth-century biology (e.g., Driesch, 1908, and his analysis of morphogenesis; Thompson, 1917, and his studies of growth and form; Waddington, 1957, and his understanding of the organism in terms of a set of potentialities). Webster & Goodwin regarded organisms (structural wholes in which the parts are to be understood in terms of their relation to each other and to their place in the overall structure) as self-organizing totalities, and morphogenesis and evolution of these organisms as systems of transformation. According to these authors, it is in the course of history that nature's hitherto unexpressed designs are gradually disclosed. In this context they quoted Waddington's conception of living organisms as devices which use the contingent 'noise' of history as a 'motor' to explore the domain of structures which are possible for them. They believed that the diversity of living forms could ultimately be accounted for in terms of a relatively small number of generative rules or laws.

The results summarized in the present essay are based on studies of the comparative morphology, systematics and evolution of Chelicerata (cf. Van der Hammen, 1979, 1982, 1985, 1985a, 1986a; see also Van der Hammen, 1986b). The choice of this group appeared to have a number of important advantages: Chelicerata are segmented animals, of which the postembryonic development is characterized by moulting and the occurrence of a series of separate instars; besides that, parthenogenesis is found in several groups of mites. The origin of segmentation (a key event in the evolution of the Metazoa) must be partly connected with a repeated manifestation of the same genotypic information;

it is often followed, in the course of evolution, by the superimposed manifestation of divergency (probably as a result of specific genetic and developmental interactions). A postembryonic development which is characterized by moulting, has the advantage that a restricted (and often fixed) number of levels of transformation can be distinguished (the manifestation of changes is concentrated at distinctly separated levels, by which an analysis is facilitated). The occurrence of parthenogenesis, finally, has the advantage that the manifestation of certain variations can be analysed not only in bisexual populations, but also in clones; in the case of the present research this analysis has led to the concept of evolutionary potentiality and the concept of probability of manifestation (cf. Van der Hammen, 1981a).

At the same time as my practical investigations in the fields of morphology, systematics and evolution, I have made studies of certain theoretical aspects of these disciplines (Van der Hammen, 1978, 1981, 1981a, 1983, 1985b, 1986, 1986c). As mentioned above, one of these studies pertained to the eighteenth- and nineteenth-century type-concept, and its origin in the Aristotelian form-concept, the Scholastic concept of substantial form and Leibniz's concepts of substantial form and monad. I subsequently developed this type-concept into a hierarchic model of the evolutionary potentialities of a group (which represents, in fact, the model of a structure in the structuralist sense).

The compendious survey of the results of my studies is divided into two parts: evolution and classification. An exposition of the first subject necessarily precedes the exposition of the second.

2. Evolution

The results of the investigations which are the subject of this section have, from the very outset, been at complete variance with the current views of evolution. For this reason, it is inevitable to start the evolutionary part of the present essay with some remarks on the theory of evolution (these remarks do not constitute an explanatory theory, but the theoretical framework necessary for a structuralist approach). The deep structure underlying evolution must be regarded as a growing structure of which the manifestation at the surface ('self-expression') has various aspects, among which 'adaptation'. This manifestation is the result of a co-ordination of internal and external factors, i.e., genotype and environment. The laws of transformation, however, are in the genotype which, consequently, is characterized by the possession of evolutionary potentialities. It may be remarked here that, in this connection, the genotype is interpreted cybernetically as a program of action and reaction. The essential elements of this theoretical framework were introduced more than a century ago by the philosopher Teichmüller (1877). It arose from his discussions with K.E. von Baer (whose evolutionary views are expounded in: Von Baer, 1876: 49–105, 170–480; and Stölzle, 1897: 195–289). Teichmüller remained un-

148

known in evolutionary biology until he was rediscovered by me some years ago (cf. Van der Hammen, 1981: 28; 1983: 187–188).

The growth of the deep structure underlying organic evolution started with the manifestation of primordial life, and progressed when living substances began to exist as definite bodies, and when reliable methods of self-replication and bisexual reproduction arose. The first important ascent in the scale of increasing complexity consisted in the origination of multicellular organisms, whilst the origin of many phyla is probably closely connected with the subsequent origin of segmentation. In the introduction, I have already mentioned the importance of a structuralist approach in the study of segmentation and the recognition of two levels: a level of similarity and a superimposed level of divergency. The origination and subsequent evolution of segmentation is an event of which a hypothetical explanation in terms of genetic control is of paramount importance; a similar explanation of Lankester's so-called laws of metamerism (see Lankester, 1902, 1904) would also be of great interest.

My structuralist approach in the study of evolution, i.e., my search for laws underlying the manifestation of evolution, is demonstrated here with the help of four completely different subjects: numerical changes in patterns of setae and setiform organs, evolutionary changes in the life-cycle, the evolution of the appendages, and transformation of form.

2.1. *Laws underlying numerical changes in patterns of setae and setiform organs*

Many details with reference to a hierarchy of systems of transformation are now known in the case of numerical changes in patterns of setae and setiform organs in Actinotrichid mites (a superorder of Chelicerata). Data with reference to these changes were recently summarized by me (Van der Hammen, 1981a; see also Van der Hammen, 1985b: 394–397).

A detailed study of the numerical variations in clones of the Oribatid mite *Platynothrus peltifer* (C.L. Koch), particularly those with reference to setae and other setiform organs (see essay II and Fig. 2.2), has revealed that there are two types of variations: anomalies and so-called vertitions. A vertition is a discontinuous individual variation (absence or presence) pertaining to an idionymous element (i.e., an element capable of receiving an individual notation) normally present, resp. absent, in specimens of the species, at the same level of postembryonic development, provided that this variation has an evolutionary significance and is originally unilateral (although there is a possibility of symmetrical manifestation). In contradistinction to mutations, vertitional changes are not immediately hereditary; the offspring inherits a certain probability of manifestation. This probability can be calculated statistically, and can be expressed in a value varying from 1 (absolute presence) to 0 (absolute absence). Vertitions generally manifest themselves bilaterally only in cases of a high probability of manifestation. Evidently, this type of variation does not pertain to a single, sudden change in the genotype, but to a system of genotypic interactions, of which the evolutionary effect manifests itself, repeatedly and to an increasing degree, in the course of time. Many so-called mutations could, in reality, belong to this type of variation.

It is now interesting to study this type of variation in various populations of one species, e.g., in the case of the number of ungues in the claws of Oribatid mites (see essay II, Fig. 2.3; the following results are similar for the cases of parthenogenetic and bisexual populations). Ancestrally, these claws are tridactyl (present three ungues) in all instars, but in most Oribatid mites they have been subject to regression, and have become monodactyl in all immature instars. Several species are known, in which the regressive evolution is still in course, and in which tridactyl, bidactyl and monodactyl legs are found simultaneously in one specimen. In some of these species, the variation in the number of adult ungues was studied in great detail, in various populations. At first sight, the manifestation of regression in the eight legs of such species appears to be whimsical: the distribution of the number of ungues in the four pairs of legs, and at the left and the right side, seems to be accidental and without any definite pattern. A statistical study, however, reveals that regression can, e.g., statistically start with the anterior legs, whilst one of the two lateral ungues can be statistically stronger (more resistant to regression) than the other. In one case, an advanced mathematical analysis revealed the existence of distinct, although very complicated connections between the variations; a number of elementary processes, pertaining to strength and priority, appeared to have acted together and in superposition, whilst there was also a certain balance (the association of opposite and mutually compensatory variations) between the two lateral ungues in neighbouring symmetrical or adjacent legs. In several cases, there appeared to be also a statistical relation between the number of ungues in certain populations and ecological or geographical data. In one case (a species of Oribatid mite from the litoral), the co-ordination of internal and external factors (co-ordinated changes in genotype and environment) was analysed in great detail (see Boelé & Van der Hammen, 1982): the statistical presence of the lateral ungues appeared to decrease gradually along the coast, from Norway to Northern France, and to increase again, suddenly, in material from Brittany (France).

The full evolutionary significance of vertitional variations can only be understood when a comparative study is made of numerous related species. Many data, e.g., are known in the case of the regressive and vertitional evolution of the so-called accessory setae of the tarsus in the Oribatid superfamily Nothroidea (see essay II, Fig. 2.5). The highest number of these setae is known from *Heminothrus targionii* (Berlese); the total disappearance is, e.g., known from the genus *Trimalaconothrus*. Numerous intermediate cases have been studied in great detail, by which it has been demonstrated that the superfamily is characterized by a general potentiality for suppression of all accessory setae, whilst this suppression manifests itself in a definite order. In various species of Oribatid mites, the probabilities of manifestation of single setae have now been studied, and the evolutionary significance of these variations (presence or absence; the probabilities of manifestation varying from 1 to 0, or the reverse, with many intermediate values) have become evident.

Because the chelicerate postembryonic development is characterized by the occurrence of instars separated by moults, the manifestation, in the course of ontogeny, of many evolutionary changes can be studied step by step. When homologous changes are studied in the postembryonic development of a great number of related species, these can be plotted side by side in a diagram, in an attempt to prepare a model of the evolution of a character in a taxon. In such a model, a so-called ontophylogenetic diagram, the ontogenetic time (t) is on the vertical axis, the phylogenetic time (T) on the horizontal axis. The ontogenies are arranged in such a way that, in the diagram, the ancestral character states are found at the left, the derived character states at the right. In simple cases, three types of ontophylogenetic diagrams can be distinguished (see essay II, Fig. 2.6):

1. The line separating, in the diagram, the ancestral character states from the derived character states (the line of chronological separation) presents, from left to right, a descendant course. An ontogeny, cutting this line, first presents the ancestral character state, then the derived character state.

2. The line of chronological separation presents, from left to right, an ascendant course. An ontogeny, cutting this line, first presents the derived character state, then the ancestral character state (ontogeny being the reverse of phylogeny).

3. The line of chronological separation is vertical, and is not cut by any ontogeny; the evolutionary change is sudden and restricted to one level (when it does not take place at this level, it does not take place at all).

More complicated diagrams (see essay II, Fig. 2.9) are, e.g., found in the case of changes which are the result of more than one regulatory mechanism, and in the case of the occurrence of highly regressive instars. Evidently, all changes in the ontogenetic aspect of numerical changes are attributable to changes in a system of genotypic interactions.

In a taxon, the resistance to regression or suppression, of the elements of a group, can present specific differences resulting in the numerical predominance of one element over another. This numerical predominance can be symbolically expressed by a priority list in which the elements are arranged in the order of decreasing strength (the order of decreasing numerical predominance and increasing weakness; see essay II, Fig. 2.11). It appears that the strongest element is not only the most common, but often also the most precocious in postembryonic development. There are indications that, besides regression, multiplication is also a sign of weakness. There are also indications that parallelisms in strength can be present in cases of metameric homonomy.

Patterns of setae appear to evolve according to very simple numerical rules. Numerical considerations generally start from primitive regular patterns in which each of the setae is individually recognizable among the other setae of the pattern (and also in the course of postembryonic development) and can be compared with the corresponding homologous seta in other specimens or species. There are some indications that these primitive regular patterns arose, in the course of evolution, from more chaotic patterns in which the setae were not yet 'individualized'. Starting from primitive regular patterns, the number of setae can decrease, in the course of evolution, by regression (i.e. suppression), or increase by multiplication of setae. Regression as well as multiplication start vertitionally and locally, and can gradually extend.

Interesting aspects of numerical evolution are revealed by the comparative developmental study of certain rows of idionymous setae, particularly in the case of regression. The postembryonic development of such rows of setae can be represented by formulae. It appears that, in the course of regression, certain stages (which are called stages of detainment) are difficult to pass, and these are represented by the most common formulae. Stages of detainment are the result of the unequal resistance of the setae to regression. Each of the common formulae (stages of detainment) is determined by the greater strength of one seta.

When, in a natural group of animals, e.g., Actinotrichid mites, the complete chaetotaxy is studied in great detail in many species, and when the postembryonic development of each of the setae is also known, it appears that the three types of evolutionary manifestation (ascendant, descendant and vertical, according to the course of the line of chronological separation in the onto-phylogenetic diagrams) generally occur simultaneously in one and the same specimen (although pertaining to different elements). It is not impossible, however, that for each of the three types the similar individual changes are correlated in some way.

Although the survey, which I have given here of the postembryonic development and the evolution of chaetotaxy, is naturally too concise, an interesting picture has yet emerged of the complicated and apparently hierarchic system of laws, which is at the base of chaetotactic changes. It has appeared, in the first place, that genotypic information pertaining to setae has an evolutionary potentiality and a probability of manifestation, and that manifestation is the result of a co-ordination of internal and external factors (genotype and environment). In the second place, it has appeared that there are, in simple cases, three types of manifestation, and that these must be connected with different regulatory mechanisms. In the third place, it has appeared that individual setae are characterized by a certain individual strength, as a result of which they are more or less liable to suppression or multiplication than other setae. Finally, it has appeared that influence is exerted by the chaetotactic pattern as a whole, as a complicated balance is maintained between the various positions.

2.2. *Laws underlying changes in the chelicerate life-cycle*

Evolutionary changes in the chelicerate life-cycle pertain to the number of successive instars as well as to the form of the instars and the types of moulting (types of moulting are not discussed in the present section; see Van der Hammen, 1980: 42, *sub* dehiscence). Several groups of Chelicerata are characterized by the occurrence of a life-cycle with a fundamental number of six instars (separated by moults); in other groups, the number of instars is greater and more irregular, whilst regression in the number of instars is also known (cf. Van der Hammen, 1978; 1985b: 397–399, Fig. 3; 1986b: 223–224). In the first-mentioned two papers, I have prepared a model of the evolution of the chelicerate life-cycle (an ontophylogenetic diagram). In this model (see essay III, Fig. 3.6), I have introduced the hypothesis that the evolution of the life-cycle has started with stasoids (instars which are morphologically different from the instar, which precedes them, by discontinuous characters, but which occur in variable numbers, so that they cannot be homologized with instars in other specimens and taxa) and presented only one immature phase, consisting of nymphoids). (A phase is a group of one or more instars, which is markedly different from preceding and following phases, and of which the instars distinctly resemble each other in cases where the phase consists of more than one instar.) It is supposed that a life-cycle with stases (instars which are, as in the case of stasoids, morphologically different from the immediately preceding instars, but which occur in fixed numbers, by which they can be homologized with similar instars in other specimens and taxa of the same group) arose from a life-cycle with stasoids, by which two evolutionary periods can be distinguished: stasoidy and orthostasy. There is much evidence that the ancestral number of stases (in orthostasy) has, in all cases, been six (protostasy; as still found in several important groups). This number could represent something like a stage of detainment in chaetotaxy, which stage is difficult to pass in the course of evolution. By different phylogenies of the levels of postembryonic development, different phases can have arisen (prelarval, larval, nymphal, adult). In the course of evolution, various levels of the ancestral protostasic life-cycle can have been subject to changes (alassostasy), including regressive evolution (protelattosis and metelattosis), the attainment of maturity at previously immature levels (neoteny), and the formation of isophena, i.e., forms that are the result of growing or repetition moults which are not associated with morphological changes except size (plethomorphosis). At a certain stage of evolution, stasoids and stases can occur in one and the same life-cycle. The possibility cannot be excluded (it is, on the contrary, rather probable) that, as to phylogeny, the model should be read in both directions and that a third evolutionary period (neostasoidy) characterized by the presence of neostasoids, and which period is still indistinguishable from stasoidy, should be recognized. It may be remarked here that groups of neostasoids (phases) can indeed be homologized with similar groups (phases) in other life-cycles. As evolutionary phenomenon, the origination of neostasoids is comparable with the origination of neotrichy, where single setae are replaced by fields (spatial patterns) of setae. In the case of neostasoidy, the instars no longer evolve as separate forms (i.e., by age), but as groups of variable numbers of successive instars (i.e., patterns in time).

The fundamental number of six stases has arisen several times in different groups of Chelicerata (parallel evolution); the potentiality of a manifestation of this number (as mentioned above, it is something like a stage of detainment) is apparently a general character of the Chelicerata (it is part of the chelicerate evolutionary program). The general laws of the numerical evolution of instars (chaotic number − fixed number − decrease or increase in number) are apparently similar to those of the numerical evolution of setae. The regressive evolution of the first instar, or instars (protelattosis; it can extend to more than one level), is a general character of all Chelicerata (it is part of the chelicerate evolutionary program). In Actinotrichid mites, other instars can also have been subject to regression (metelatosis), viz., the deutonymph in Acaridida, and the proto- and tritonymph in Trombidina; in these cases, the evolutionary phenomena are manifestations of the evolutionary programs of groups of lower taxonomic rank.

2.3. *Laws underlying the evolution of the chelicerate appendages*

The chelicerate appendages and their evolution have repeatedly been the subject of my investigations (cf. Van der Hammen, 1977; 1977a; 1985: 50−51; 1985a: 139−141, Figs. 2−6; 1985b: 400−403, Fig. 4; 1986: 24−31, Figs. 5, 7−8; 1986b: 218−220, Fig. 1). My insights in this field gradually developed, and not until recently (Van der Hammen, 1986a: 30−45, Figs. 15−23) have I succeeded in solving the last important problem (that of the homology of epimera and coxae). It has now become possible to give a more or less complete (and partly hypothetical) survey of the laws underlying the evolution of these appendages.

Chelicerata (with the exception of Xiphosura) differ from other Arthropods by the fact that the main promotor-remotor movement is at the coxa-trochanter or body-trochanter joint, and not at the body-coxa joint. It has now become evident that coxae have evolved, relatively late, from epimera (the sternites of the prosoma) by a gradual separation of these from the sternal exoskeleton, a gradual increase in movability, and a subsequent extension in antiaxial direction (associated with the inclusion of pleural regions). Epimera are present in Palpigradi, Actinotrichida, Solifugae and Pseudoscorpionida. The coxae of Arachnida s. str. (Uropygi, Amblypygi, Araneida) demonstrate the gradual transition from coxae with ventral position and slight movability to coxae with lateral position and increased movability. The problem of the homologization of leg segments (podomeres) is now definitely solved: the main promotor-remotor movement is (with the exception of Xiphosura) at the base of the trochanter; a levator-depressor movement is at the pivot joint between trochanter and femur (in case this joint is repeated, there is a second trochanter); a knee with a hinge joint is present between femur and patella (genu) or tibia (in the case of a second femur, an extra hinge joint is present between the trochanter-femur- and the knee-joint); patella, tibia and tarsus can easily be recognized because of the number of true (eudesmatic) articulations in the descending part of the leg (the articulation between basi- and telotarsus, if present, is adesmatic). This original pattern can be changed by secondary transformations (the origination of rocking joints from pivot joints, the origination of extensor muscles in hinge joints, etc.), but generally remains recognizable. The following model of the evolution of the chelicerate appendages has now been prepared by me (cf. Van der Hammen, 1985: 50−51; 1985b: 400−403, Figs. 4−5; see also essay IV, Fig. 4.4).

Because the six pairs of chelicerate appendages (chelicera, palp and legs) are homonomous structures which have developed by parallel evolution from lobopodia, the ancestral number of the segments of the chelicera (three) could be the prototypal number for all the appendages (archepodia; the deepest level of the structure underlying the evolution of the chelicerate appendages is apparently common to all the appendages); these three segments are usually called trochanter, body (of the chelicera) and apotele (or claw-segment). This number of segments could have occurred in short-legged species which stood up on their legs. In chelicerate embryos, palp and legs first divide into four segments: prototrochanter, protofemur, prototibia and prototarsus. A leg with these four segments and an apotele or claw-segment (i.e., five segments: a so-called protopodium) could represent the first stage in the evolution of a hanging stance (this second level of the structure underlying the evolution of the chelicerate appendages is apparently common to five pairs of appendages). All types of chelicerate palps and legs can be derived from this primitive type. As mentioned above, free coxae arose from epimera, and can include pleural regions. Protopodia arose from archepodia, either by a further subdivision of segments, or by the addition of projecting parts of the body (cf. Van der Hammen, 1985: 50). Because trochanter 2, in Anactinotrichida, disappears by suppression, it must originally have evolved by repeated manifestation of the same genotypic information. Because femur 2, in Actinotrichida, disappears by fusion of femur 1 and femur 2 (cf. Van der Hammen, 1982: 38; fusion takes place by the regression and disappearance of the muscles, followed by the gradual disappearance of the articulation), it must originally have evolved by subdivision of the femur. Patella (genu) and tibia originally evolved by subdivision of the prototibia; it is interesting that this subdivision has manifested itself in two different ways: the patella-tibia articulation is, ancestrally, either a hinge joint or a pivot joint. It is also interesting that a second femur arose in Epimerata (Palpigradi and Actinotrichida) and Apatellata (Solifugae and Pseudoscorpionida) only; its presence must be regarded as a unique shared derived character of these two groups (which are now regarded by me as constituting a separate class). A second trochanter is found in various groups of Chelicerata; its presence is apparently founded on a general evolutionary potentiality of the Chelicerata. It is quite understandable that repetition of information, in the case of trochanter 2, is a shared evolutionary potentiality of all Chelicerata, but that division of the femur (the development of new characters) is a shared derived character of some groups only.

The data mentioned in this section demonstrate that it has been possible to reconstruct the general laws underlying the evolution of the chelicerate appendages, as well as the hierarchy of evolutionary programs. It may, finally, be remarked here that, from a certain moment in evolution (probably coinciding with the beginning of terrestrial life), there apparently arose a tendency towards a decrease in the number of segments in the ascending part of the legs, and a tendency towards an increase in the descending part.

2.4. *Laws underlying transformation of form*

Three-dimensional changes of form are more difficult to analyse than simple numerical changes. These can, however, also be understood as the result of changes in a system of genotypic interactions and the changes of growth-rates involved in it. When Darwin was still alife, Von Baer (cf. Von Baer, 1876: 49–105, 170–480; Stölzle, 1897: 195–289) gave already as his view that transmutation is a process of change manifesting itself in the course of ontogeny and resulting in important (i.e., discontinuous) modifications. (Modifications are, generally, more important as they manifest themselves earlier in embryonic development.) Since D'Arcy Thompson (1917) we know that, in a taxon, differences in form (even important ones) can be explained in terms of relative growth (he analysed transformation of form by the method of co-ordinates). Needham (1936) remarked that nobody had applied D'Arcy Thompson's method to embryology and morphogenesis, whilst growth-rate and differentiation-rate yet seem to be aligned with one another.

Considerable progress could already be made when differences in form were analysed in much greater detail in a single taxon, e.g., in all representatives of a genus, so that all transformations could be understood as resulting from changing growth rates.[1] Particularly in the case of Chelicerata, changes could then be studied in the postembryonic development and in different populations, in order to establish if discontinuous variations with an evolutionary sense are present. This type of research will certainly be time-consuming, but the results will probably be of great value: laws underlying transformation of form could, in this way, be disclosed.

---

[1] An analysis of this kind was carried out by Gould (1984) for the case of shells of the Gastropod genus *Cerion*.

156

2.5 Conclusion

Although the structuralist approach in the study of evolution has been demonstrated with the help of four groups of phenomena only, it is evident that, in this way, general laws underlying the manifestation of evolution can be disclosed, and that these laws pertain to hierarchic systems of interactions and the gradual manifestation of evolutionary potentialities. Each level of the hierarchic system of interactions represents the realization of potentialities and includes new potentialities (the development of new potentialities, however, is an important unsolved problem). Taxa are apparently characterized by evolutionary programs of which generally, at a certain moment of evolution, not all parts have become manifest (manifestation is supposed here to be regulated cybernetically). It has also been demonstrated that evolutionary changes can be represented schematically by diagrams, etc. Evidently, it is possible to represent all the evolutionary potentialities of a taxon by a series of similar diagrams. A complete series of these diagrams, hierarchically arranged, will represent what I have called the archetype of a group (Van der Hammen, 1981): the complete representation of its evolutionary potentialities and their modes of manifestation. The construction of such an archetype (a model of the deep structure) is within our reach, although it presupposes a wide field of research. Realization of this project will again be time-consuming, but I am convinced that, after the construction of a model of the deep structure of a group, and after the mathematical analysis of the complicated correlations, etc., something of the process of evolution will at last be understood.

3. Classification

From a structuralist point of view, data supplied by the study of the overall similarity of representatives of a taxon, constitute a surface structure which is a reflection of relations at the levels of deeper structures. Whilst a phenetic classification is evidently based on a surface structure, evolutionary and phylogenetic classifications should be based on the deep structure (a hierarchy of synchronic systems). A model of this hierarchy is constituted by the archetype as defined in the previous section. It will be evident that, once the archetype has been constructed, classifications based on it will be easily comparable and even more or less similar. Brooks & Wiley (1985) recently argued that the deep structure of systematic techniques is constituted by systematic theory. This is not a very fruitful idea, because theory is a man-made structure which has no direct relationship to the naturalness of a classification (evolutionary or phylogenetic).

In the previous section, it was pointed out that the hierarchy of deep structures is characterized by the presence of evolutionary potentialities which can manifest themselves separately (and discontinuously) by parallel evolution,

I repeat here that parallel evolution (cf. Van der Hammen, 1986) should be regarded as the separate manifestation, at the level of the surface structure, of identities at the level of the deep structure. The important difference between the evolutionary potentialities of the deep structure and their manifestation at the surface, is generally not taken into account in current work in the field of phylogenetic systematics. It is now evident that branching points in a scheme underlying phylogenetic classification should correspond with the separation of systems of transformation and evolutionary programs, and not with the manifestation of their evolutionary potentialities at the level of the surface structure (the distinction between the development of new potentialities, on the one hand, and changes in systems of interactions, on the other, should constitute one of the foundations of taxonomy). It is also evident that the hierarchy of the natural classification should correspond with the hierarchy of deep structures. In the previous chapter, it has been demonstrated that evolutionary changes in chaetotaxy, life-cycle and segmentation of the legs, can be attributed to definite levels in the hierarchy of systems of transformation and, as a consequence of this, to definite levels in the hierarchy of the natural system. The structuralist approach (it could be characterized as structuralist taxonomy) is particularly important in the case of higher classification, and it decisively influenced, indeed, my final view of the higher classification of Chelicerata. Especially at the supraordinal levels, current taxonomic methods tend to develop atomistic views in which the organism as a functional whole has virtually no place, and in which the hierarchic value of characters is not sufficiently analysed and discussed. It is evident that, pre-eminently in this field, the structuralist approach will advance the naturalness of a classification.

References

Baer, K.E. von, 1876. Reden gehalten in wissenschaftlichen Versammlungen und kleinere Aufsätze vermischten Inhalts, 2. Studien aus dem Gebiete der Naturwissenschaften. – St. Petersburg: xxvi + 408 p., 22 figs.

Boelé, F.E. & L. van der Hammen, 1982. Variation in the number of ungues in Ameronothrus schneideri (Oudemans), an Oribatid mite. – Zool. Meded. Leiden 56: 169–191, Fig. 1, Tab. 1–5.

Brooks, D.R. & E.O. Wiley, 1985. Theories and methods in different approaches to phylogenetic systematics. – Cladistics 1: 1–11.

Driesch, H., 1908. The science and philosophy of the organism. – London (Adam & Black). Vol. 1: xiii + 329 p.; vol. 2: xvi + 381 p. [I have used the fourth (abridged and revised) German edition, entitled: Philosophie des Organischen (Leipzig, 1928)].

Gould, S.J., 1984. Morphological channeling by structural constraint: convergence in styles of dwarfing and gigantism in Cerion, with a description of two new fossil species and a report on the discovery of the largest Cerion. – Paleobiology 10: 172–194, Figs. 1–4, Tab. 1–8.

Hammen, L. van der, 1977. A new classification of Chelicerata. – Zool. Meded. Leiden 51: 307–319, Fig. 1, Tab. 1–3.

Hammen, L. van der, 1977a. The evolution of the coxa in mites and other groups of Chelicerata. – Acarologia 19: 12–19, Figs. 1–2.

158

Hammen, L. van der, 1978. The evolution of the chelicerate life-cycle. – Acta Biotheoretica 27: 44–60, Figs. 1–6. [Cf. essay III.]

Hammen, L. van der, 1979. Comparative studies in Chelicerata I. The Cryptognomae (Ricinulei, Architarbi and Anactinotrichida). – Zool. Verh. Leiden 174: 1–62, Figs. 1–31.

Hammen, L. van der, 1980. Glossary of acarological terminology, 1. General terminology. – The Hague (Dr W. Junk Publishers): vii + 244 p.

Hammen, L. van der, 1981. Type-concept, higher classification and evolution. – Acta Biotheoretica 30: 3–48, Figs. 1–7. [Cf. essay V.]

Hammen, L. van der, 1981a. Numerical changes and evolution in Actinotrichid mites (Chelicerata). – Zool. Verh. Leiden 182: 1–48, Figs. 1–13. [Cf. essay II.]

Hammen, L. van der, 1982. Comparative studies in Chelicerata II. Epimerata (Palpigradi and Actinotrichida). – Zool. Verh. Leiden 196: 1–70, Figs. 1–31.

Hammen, L. van der, 1983. Unfoldment and manifestation. The natural philosophy of evolution. – Acta Biotheoretica 32: 179–193. [Cf. essay I.]

Hammen, L. van der, 1985. Comparative studies in Chelicerata III. Opilionida. – Zool. Verh. Leiden 220: 1–60, Figs. 1–34.

Hammen, L. van der, 1985a. Functional morphology and affinities of extant Chelicerata in evolutionary perspective. – Trans. Roy. Soc. Edinburgh, Earth Sciences 76: 137–146, Figs. 1–8.

Hammen, L. van der, 1985b. A structuralist approach in the study of evolution and classification. – Zool. Meded. Leiden 59: 391–409, Figs. 1–5.

Hammen, L. van der, 1986. On some aspects of parallel evolution in Chelicerata. – Acta Biotheoretica 35: 15–37, Figs. 1–9. [Cf. essay IV.]

Hammen, L. van der, 1986a. Comparative studies in Chelicerata IV. Apatellata, Arachnida, Scorpionida, Xiphosura. – Zool. Verh. Leiden 226: 1–52, Figs. 1–23.

Hammen, L. van der, 1986b. Acarological and arachnological notes. – Zool. Meded. Leiden 60: 217–230, Figs. 1–3.

Hammen, L. van der, 1986c. Some notes on taxonomic methodology. – Zool. Meded. Leiden 60: 231–256. [Cf. essay VII.]

Hammen, L. van der, in press. Structuralism in evolutionary biology and systematics. – Rivista di Biologia 80(3). [Abstract in: Rivista di Biologia 80(2): 216–220.]

Lane, M. (ed.), 1970. Structuralism: A reader. – London (Jonathan Cape): 456 p.

Lankester, E.R., 1902. Arthropoda. – Encyclopaedia Britannica (tenth edition) 25: 689–701, 11 figs.

Lankester, E.R., 1904. The structure and classification of the Arthropoda. – Quart. Journ. Micr. Sci. (N.S.) 47: 523–582, Pl. 42.

Needham, J., 1936. Order and life – Yale University Press. [I have used the paperback edition, Cambridge, Mass. & London (The M.I.T. Press), 1968: xxii + 175 p., Figs. 1–45.]

Piaget, J., 1968. Le structuralisme. – Paris (Presses Universitaires de France): 127 p. [English translation, by C. Maschler, entitled Structuralism, London (Routledge and Kegan Paul), 1971: 153 p.]

Stölzle, R., 1897. Karl Ernst von Baer und seine Weltanschauung. – Regenburg: xii + 688 p.

Teichmüller, G., 1877. Darwinismus und Philosophie. – Dorpat: viii + 90 p.

Thompson, D'A.W., 1917. On growth and form. – Cambridge (University Press): xvi + 793 p. [I have used the 1977 reprint of the abridged paperback edition, edited by J.T. Bonner (Cambridge University Press). In this edition, the chapter On the Theory of Transformation, or the Comparison of related Forms is found on p. 268–325.]

Waddington, C.H., 1957. The strategy of the genes. A discussion of some aspects of theoretical biology. – London (Allen & Unwin): ix + 262 p., 37 + X figs.

Webster, G. & B.C. Goodwin, 1981. History and structure in biology. – Perspectives in Biology and Medicine, Autumn 1981: 39–62.

Webster, G. & B.C. Goodwin, 1982. The origin of species: a structuralist approach. – Journ. Soc. Biol. Struct. 5: 15–47.

VII. SOME NOTES ON TAXONOMIC METHODOLOGY

1. Introduction

In the previous essays, especially in essays V and VI, a structuralist approach in the study of systematics has been developed, which can be particularly useful in higher classification. This structuralist method is supplemented here with a general introduction to taxonomic methodology.

The first draft of the present essay was written many years ago as an introduction to methodology for private use. Because my own systematic and morphological studies are in the field of Chelicerata, many subjects were treated with this group in mind, and most of the examples were chosen from this group as well. The essay was first published (Van der Hammen, 1986c), entirely revised and partly rewritten, as the final part of a series of theoretical papers, in which my views with respect to evolution and classification are expounded (see also Van der Hammen, 1978a, 1981, 1981a, 1983, 1985a, 1986). This series of theoretical papers was written in the same period as the four parts of my series of Comparative Studies in Chelicerata (Van der Hammen, 1979a, 1982, 1985, 1986a) and, in writing the two series, insights developed and influenced each other mutually. The essay is now revised and adapted to the purpose of the present book.

The origin of biological systematics is found in folk taxonomy, which includes folk nomenclature and folk classification, and from which it grew away in the course of time. Folk nomenclature can be simple, complex or composite (in the latter case sometimes even binominal). Folk classification is often hierarchical (a category at one level is included in a category at the next higher level); taxa at the same level are differentiated by contrast. Criteria for classification can be morphological as well as functional (the use as food, medicine, ornament, etc.) (see Conklin, 1962).

The zoology of Aristotle, the father of scientific taxonomy, was founded on folk taxonomy. Aristotle dealt with species and groups, and his names originated for the greater part from folk nomenclature (he extended, however, the number of groups, and introduced several new names) (see Aubert & Wimmer, 1868: 55–184). For each of his groups, Aristotle generally described the characters in common and the differences with other taxa. Several of the groups distinguished by him exactly correspond with our modern views of

160

them. Aristotle mentions two categories: *genos* and *eidos*. In his biological work, the terms are not always unequivocally distinguished (more clearly, however, in his logic). Particularly in the introductory parts, the term *genos* (which is used more frequently) pertains to groups; in this case, *eidos* is subordinated and can pertain to species (see Balme, 1962). Aristotle arranged his groups in a descending scale of nature, extending from man to inanimate nature.

The subsequent development of systematics, particularly from the sixteenth and seventeenth century onward, included an increase in the number of known species (also as a result of the exploration of newly discovered parts of the world), the consistent application of the binominal method of nomenclature, the development of a rigid hierarchy of taxonomic categories, and the increased emphasis on classification by overall similarity (to replace the delimitation of taxa by a few key-characters only).

An important part of systematics had already been accomplished according to intuitive methods, by specialists who generally had not received any instruction in taxonomic theory and philosophy of science, when an increased interest arose in its methods and principles. This is indicated by the foundation of taxonomic journals and societies, and by the appearance of a large mass of theoretical publications, among which the books by Hennig (1950, 1966), Remane (1952, 1956), Mayr, Linsley & Usinger (1953), Mayr (1969), Simpson (1961), Sokal & Sneath (1963), Sneath & Sokal (1973), and Wiley (1981), for instance, constituted important landmarks. This interest in taxonomic methods and principles led to a perfection of phenetic classification by numerical procedures, to a perfection of evolutionary classification by phyletic weighting, and to a perfection of phylogenetic classification by the study of evolutionary branching sequences.

Philosophy of science is a rational reconstruction and a logical analysis of scientific activity, particularly of the intuitive practice of leading scientific investigators. Its search is for a justification of scientific concepts, laws and theories. In the present paper, an attempt is made to reconstruct some of the basic principles and methods of biological systematics (particularly systematic zoology). It contains an analysis of systematic practice, a survey of the kinds of attributes, and further sections on observation, methodology, comparative study, the species-concept, and biological classification.

2. Analysis of taxonomic practice

Systematics, as a branch of biology, is a science devoted to the distinguishing, naming, describing and ordering of organisms; zoological systematics is restricted to the animal kingdom. A first general view of the field of activity of a systematic zoologist (we must admit that, even as a beginner, he starts already with a large body of information, prejudices, traditions, literature, etc., which greatly influence his observations and conclusions) can be obtained by analysing his successive performances in any particular case, for instance, in studying a sample of the soil-fauna obtained by extraction from litter and moss by the use of a so-called Berlese funnel. Such a sample can consist of a multitude of specimens arranged just anyhow, in chaotic disorder. Each of these specimens presents numerous individual attributes, observable by means of the senses and with the aid of optical instruments. Consequently, the body of existing information, etc., the attributes of the specimens in the material, and attentive observation, constitute the starting-point of the systematist's scientific activity. The systematist's next step, in the study of his chaotic material, is comparison: he considers various specimens and their attributes in connection with each other, and marks the similarities and differences. In order to convey the results of the comparison in language, a terminology is required; this can partly be adopted from existing traditions in the field of systematics, and from auxiliary sciences, and be supplemented, if necessary, by new terms. After a comparison of the specimens, the next step can be a simple sorting: specimens can be arranged in groups according to similarities and differences in shape, by which procedure the occurrence of a certain number of different forms can be established. A first difficulty is constituted by the fact that the relations of certain different forms (male, female, immature forms) to each other, can at first be unknown. The term phenon, which was introduced by Camp & Gilly (1943: 335–337) for a botanical sample that is phenotypically homogeneous, can be applied to the different forms in the material. It may be remarked, however, that Sneath & Sokal (1962: 859–860; 1973: 294–295) used the term in a different sense, and applied it to any group established by numerical taxonomy; phenon, in the present meaning of the term (see also Mayr, 1969: 5), is more or less synonymous with Hennig's semaphoront (Hennig, 1950: 9; 1966: 6). Different phena of one species can be associated with each other by comparison with data from literature, by a study of the reproductive behaviour, and by breeding. Not until a comprehension of the connections between phena, can we arrive at a full recognition of species, the fundamental units of systematics.

With the data available to us (e.g., from literature), the species in the material can be identified: they are recognized as belonging to certain units, and as different from other units. In order to put the identification into words, the units should bear a name. For this reason, a special branch of systematics is constituted by nomenclature (which is regulated by rules).

Sorting, comparing, identifying and naming belong to the first operations of the systematist. These operations can be completed by a description of the material, and, if necessary, by a further analysis of it.

The next step of the systematist is constituted by a further arrangement of the material (in this case, a more extensive material than that of our sample is generally required). Different species are placed close to, or far from, each other. This is measured by the degree of overall similarity (based on the number of attributes in common). This results in a so-called phenetic arrangement

162

which, generally, is largely based on the external morphology. A phenetic arrangement can be the starting-point for grouping, i.e., the placing of similar species in higher units, separated from groups of the same kind by distinct discontinuities. Similar operations can be repeated at higher levels. All of these operations belong to the field of classification, and pertain to the delimitation, ordering and ranking of species and groups (genera, families, orders, classes, etc.). The final presentation of the classification, however, depends on the ideal one has in view.

All operations described above, in our analysis of taxonomic practice, belong to the field of activity of the systematist. As a result of this analysis, the definition of systematics given at the beginning of this section can now be slightly extended in the following way: systematics is a branch of biology, devoted to the distinguishing, naming and description of organisms, and to the arrangement and classification of these, in one way or another, according to similarity and relationship.

Besides systematics, the term taxonomy is also in use. The word taxonomy was introduced by the French botanist De Candolle (1813: 23–25) and pertained to the theory of classification. Simpson (1961: 11) defined taxonomy as the theoretical study of classification, including its bases, principles, procedures, and rules. According to this definition, the subject of the present paper (taxonomic methodology) is taxonomy; it constitutes an indispensable branch of systematics.

3. Kinds of attributes

As mentioned above, in section 2 of the present essay, the attributes of the material constitute one of the starting points of investigations in the field of biological systematics. The organisms which constitute the material usually present numerous attributes, among which morphological, cytological, histological, genetic, ontogenetic, physiological, biochemical, ecological, geographic, and ethological attributes. All kinds of attributes can be included in a systematic study, although many systematists often confine themselves to morphological attributes. It has even been suggested that the execution of a function can only be understood by a study of the form (Kälin's logical primacy of morphology; see Kälin, 1941: 15). Many of the attributes of the material will indeed be correlated, but this is of no importance for a first comparative study in which all attributes have the same value. Not until a later stage of the study, will the attributes be evaluated.

Taxonomic methods and principles are the same for all kinds of attributes. We can hardly speak of a new systematics when we introduce new kinds of attributes. Systematics is only fundamentally changed by the introduction of new methods.

In the present section, various kinds of attributes will be briefly discussed, and a few examples (mainly from Chelicerates) will be given as illustration.

An important part of systematics is based on morphological attributes, because these can still be studied in preserved material. Most attention is generally paid to external morphology, although internal anatomy can be equally important (internal anatomy often represents a more generalized and conservative condition, and can be particularly important at the higher levels of classification). Emphasis is often laid on certain aspects of morphology, such as the exoskeleton and the genitalia in Arthropods. Chaetotaxy (and phanerotaxy in general), i.e., number and arrangement of setae (and setae-like organs) on body and appendages, is of particular importance in the study of Chelicerata.

Systematics can also make use of attributes from the fields of cytology and histology. Onychophora, e.g., are characterized by the presence of unstriated muscles, Myriapoda and Hexapoda by the presence of striated muscles. Modern systematics attaches much value to sperm morphology and spermiocytogenesis.

Attributes from cytogenetics (a branch of genetics dealing with its cytological aspect), used by systematists, include chromosome number and shape, and types of nuclear division and sex determination. Attributes from formal genetics, used by systematists, include the effects of hybridization (fertility points to close relationship).

Variability is of particular interest to systematics. Before closely related species can be separated, the range of variation of several attributes must be closely studied. Generally, a systematist describes also the anomalies in his material.

Ontogeny is a dynamic process, and attributes from ontogeny must, consequently, pertain to change. As soon as we isolate, from ontogeny, a static moment (e.g., the structure of a larva), the attributes are morphological. In many groups, ontogeny can be subdivided into an embryonic and a postembryonic period. The inversion of the curvature of the germ band in various groups of Spiders, e.g., constitutes an attribute from the embryonic period. Ontogenetic attributes of considerable systematic importance are those pertaining to the postembryonic development, such as the numbers and kinds of instars in Chelicerata, the types of moulting in Oribatid mites, the addition of segments in the paraproctal region of Actinotrichid mites, and the development of the number of setae and other phaneres in Actinotrichid mites (this development can be represented symbolically by formulae).

Attributes from the field of morphogenesis include regeneration, i.e., the renewal of a portion of body or appendages, which has been completely or partly lost. In mites, e.g., lost legs (or parts of legs) can be regenerated, in the course of moulting, in various characteristic ways.

Physiological attributes are rarely considered by systematists, although these can be of great interest. I point out the occurrence of either internal or external digestion in various groups of Chelicerata, and to the efficiency in oxygen supply in various groups of Spiders.

Attributes from biochemistry and serology include the presence (in different

164

quantities) or absence of relatively simple substances, the information content of highly complex structures (such as the nucleic acid coding of the genome, and protein sequences), and the production of antibodies in experimental animals (serology).

Ecological attributes, of interest to systematists, include habitat, food preference, tolerance to various physical factors, resistance to predators, parasites, and host preference. Among Myriapoda, Centipedes are carnivorous, whilst Millipedes generally prefer food of vegetable origin. In the Oribatid mite family Ameronothridae, most species (as in the related families Fortuyniidae and Selenoribatidae) are inhabitants of the litoral, although there is one group of inland species (a representative of the related family Podacaridae is also terrestrial). In the Oribatid family Zetorchestidae, species of three genera (*Saxicolestes, Belorchestes* and *Litholestes*) are found on bare rocks, where they feed on pollen grains and the like, transported by air; Zetorchestid species of other genera are found in moss and litter. Larvae of Trombidei (Actinotrichid mites) are generally parasitic; larvae of the family Trombiculidae nearly exclusively parasitize Vertebrates, those of the family Trombidiidae Arthropods. Many parasites are host-specific (and the reverse: many hosts have specific parasites).

Geographical distribution also belongs to the attributes of subspecies, species and higher units. Among the Solifugae (a group of Chelicerata), two families (Eremobatidae and Ammotrichidae) are found in America, whilst the remaining families are all found in the Old World. Among the Ricinulei (another group of Chelicerata), one genus (*Ricinoides*) is West African, whilst the remaining genera occur in the eastern part of South America, in central America, Mexico and South Texas. Gertsch (1964) mentioned four species of the North American Spider genus *Hypochilus*; of these, *H. thorelli* Marx is found in the southeastern United States, *H. gertschi* Hoffmann in Virginia and southern West Virginia, *H. bonneti* Gertsch in Colorado, and *H. petrunkevitchi* Gertsch in California. According to Vachon (1952: 248−255), two subspecies of the Scorpion *Buthus atlantis* Pocock are found in Morocco; *B. atlantis atlantis* and *B. atlantis parroti* Vachon. The typical subspecies is found on the Atlantic coast (in the sand), near Mogador and Agadir; the subspecies *parroti* is found in the Sous valley, in Argan forests (at a distance of about 40 km from the coast).

Many attributes from ethology are of taxonomic interest; these include mechanisms of various complexity. Simple behavioural patterns can be integrated into more complex ones. Nest-building in birds is based on a coordination of various patterns. Valuable attributes are constituted by animal sounds (Mammals, Birds, Amphibia, Insects). In complex behavioural patterns, several systematic levels can be distinguished. The web of the Spider *Zygiella x-notata* (Clerck), which is the result of a complex behavioural pattern, is a typical orb-web of the family Argiopidae; as in other species of the genus *Zygiella*, one segment of the orb is missing, and a line in the middle of this region leads from the centre to the Spider's retreat. The web of *Z. x-notata* differs, however, from that of other species of the genus, by the general size, by the number of radii, by the attributes of the hub, and by the size of the so-called free zone.

In the above, a few examples are given of the various kinds of attributes, which the systematist can include in his study. They illustrate the variety and the numerousness of the attributes from which the systematist must select the useful characters. Many of these attributes can easily be studied by himself. For other attributes he will need the assistance of specialists in other branches of biology. More than ever before, systematics has become a branch of biology with various auxiliary sciences, and with many possibilities of co-operation.

4. Observation

The fundamental data which constitute the base of systematics are obtained, as in other sciences, by observation. Consequently, a brief discussion of observation must be one of the starting-points (besides the attributes of the material) of an introduction to the foundations of taxonomic theory. Although many technical aids have been developed to improve and extend our observations, the present discussion starts from sensory perception which constitutes the base of our relations to objects. It must be repeated here, however, that every scientific observer, even a beginner, starts his observations with a large body of information, which greatly influences the results of the act of observing.

From the view-point of sensory physiology and psychology, observation pertains to stimuli from objects, which affect the senses, are transported to the brain, and become conscious as sensations which are then referred to the external object by the action of the mind. In the act of observing, memory and the imaginative faculty play a prominent part; the process is, however, experienced in its totality. Every perception can be subject to unconscious corrections before it is consciously experienced. Perceptions are, moreover, accompanied by an emotional response. Observation not only has physiological and psychological, but also philosophical (particularly epistemological) aspects; observation is one of the sources of our knowledge. Evidently, a deeper understanding of observation (and its reliability) is indispensible for an evaluation of the knowledge acquired by it.

That which is consciously perceived never exactly corresponds with its external object: our senses omit certain details, add something, and distort other things. Besides that, the threshold of our powers of observation is liable to fluctuations. Factors like expectation, attention and concentration play also an important part. Fluctuations of the attention depend on our state of health, the degree of tiredness, interest and personal concern. The development of an observation program, and the recommencement of certain difficult series of observations, favourably influence the powers of observation.

A critical observer must have a certain knowledge of the various aspects of observation (and of the instruments used by him to extend and improve his observations). Besides that, a carefully defined vocabulary is required for an unambiguous description of details perceived. The greater part of the attributes observed are in the visual field; in several animal groups, sounds and/or smell constitute also interesting attributes. It may be remarked here that I have written a series of papers on the observation of nature, in which touch, smell, taste, sound, form, colour, light and dark, and motion, are dealt with separately (Van der Hammen, 1971, 1972, 1974, 1975, 1975a, 1975b, 1977, 1978, 1979). Much attention is paid in these papers to the development of a standard vocabulary and a standard formula of description. A revised edition in book-form (the original papers are in Dutch), with a theoretical introduction, is in course of preparation.

5. Methodology

By attentive observation of the attributes of the material, and influenced by information already existent, the systematist collects the facts required for his study. These facts are considered then in relation to each other, and hazy notions arise about characters in common. Facts are ordered by the formation of concepts. These concepts can, subsequently, be indicated by a name or term, and the systematist can try to define the term as detailed as possible. A definition must be a precise and unequivocal explanation of the meaning of a term. There are various kinds of definitions and various rules for defining, which are not dealt with here. Many dictionaries of general biological terms are now available, and many glossaries of the terminology of particular groups.

Starting from the facts, and by the description of these (made possible by concepts, terms and definitions), the systematist, in distinguishing species and preparing classifications, follows (often unconsciously) a definite way. This way leads from the facts, through the construction of hypotheses and the deduction of predictions, to testing (evaluating in the light of newly discovered facts). A systematist describing, for instance, a new species, is usually not conscious that this procedure includes the introduction of several hypotheses, based on a restricted material, to be tested in the light of new specimens. By hypothesis we understand a supposition, a provisional answer to a problem. Evidently, the nature of the problem, and the aim of the investigation must be known, and the construction of hypotheses must be preceded by a formulation of the problem.

Besides hypothesis, mention must be made here of theory (the theory of evolution, for instance). A theory is a system of concepts, laws, hypotheses, etc., held as an explanation of phenomena. It can be philosophical rather than scientific, and constitute a way of looking at nature, guiding our perceptions. A natural law is a general statement about a regular connection between facts from nature; a rule pertains to that which is normally (but not always) the case.

In the following subsections, some of the systematist's methods will be discussed in more detail.

5.1. *Description*

A description is a representation in language, symbols and illustrations, of the attributes observed during the study of an organism; as many aspects of the object as possible must be considered for description.

In order to be able to prepare a workable description, the investigator must have developed a profound knowledge of the group to which the object in question belongs, and of all aspects involved. If possible, a representative sample should be studied, in order not to neglect the variability aspect.

The aim of a description is to collect and record data for subsequent recognition, and for comparison with other data. In order to prepare a description which is characteristic, the investigator must select attributes, although his choice should not be restricted to differential characters (which are the object of a diagnosis). By the necessary selection of attributes, the description contains a subjective element (often as the result of a tradition), although we can aim at a certain objectiveness by carefully explaining the choice.

In order to be easily comparable, a description must conform to a certain sequence, and to a standard orientation of the illustrated structures. A description is more easily recognized by other investigators, when special attention is paid to linguistic usage and to the preparation of clear illustrations. It must be borne in mind that, in a description, an investigator gives evidence of his powers of observation, as well as of his ability to reproduce the attributes observed.

5.2. *Comparison*

Comparison (see section 6) is the action of putting two or more specimens, species or groups side by side, and of considering them in connection with each other, in order to mark the similarities and differences. Many of the conclusions at which a systematist arrives, in the course of his taxonomic study, are based on comparison. As in the case of a description, a comparison presupposes a profound knowledge of the material and its attributes. In the course of a pure comparison, as many attributes as possible are compared, and no restrictions as to what will be compared are made beforehand; during a first comparison, all attributes have equal value. In the case of a particular attribute, comparison can pertain to similarities or differences, either in degree or in numbers. In many cases, the objectivity of a comparison can be improved when similarities and differences are measured; the computation of a measure of resemblance can be done in various ways (see Sneath & Sokal, 1973).

5.3. *Arrangement and classification*

Arrangement is the action of placing forms, on their resemblance or difference (as a result of comparison), near to or far from each other; resemblance can pertain, in this case, to the degree of overall similarity or to particular selected attributes. When the degree of resemblance is computed, arranging has the meaning of measuring.

Classifying (see section 8) is the action of distributing forms, which are placed near to or far from each other, in discrete classes on the ground of particular identical or similar attributes. A classification, consequently, is characterized by a principle of division.

Classification is the action of classifying, as well as the result of it. In both senses, its aim is a conversion of chaos into order, and the preparation of a general view of a complicated multitude of forms. By classification, the knowledge acquired in the course of time is organized and easily available, whilst the inclusion of new knowledge is facilitated. A classification, moreover, facilitates discovering new ways of study.

Although a collection of objects can be classified in various ways, one should search for the least artificial, the most natural classification (which takes into account the greatest possible number of similarities and differences); in this way, the greatest possible numbers of relations (proximities and distances, affinities or relationships) between the classified objects become evident.

The characters used in a classification are of different kinds; one can distinguish, for instance, quantitative and qualitative characters (the first-mentioned characters pertain to numbers and measurements). Some characters present gradations, from complete absence to maximum development. These gradations can be quantified, because total absence can be represented by 0, maximum development by 1 (or 100).

In classifying, philosophical methodology distinguishes between genus and species; genus refers to any group comprising two or more species, species to any group which forms part of a larger group. Biological systematics makes use of these terms in a particular way by applying them to special restricted cases (the biologist recognizes, moreover, the occurrence of monotypic genera).

Various kinds of classification can be distinguished. In primitive societies (and among uneducated people), an intuitive classification of animals and plants (folk taxonomy) is found, based on unformulated experience. Generally, two types of classification are distinguished: a natural and an artificial classification. An artificial classification is based on one or a few easily visible characters. A natural classification is based on the examination of a great number of characters (as many as possible) and on the analysis of these; it is generally conceived as a classification based on the hypothesis of genetic relationship and common ancestry. In the case of natural classifications, various models of classification are possible. In the hierarchical model, each group forms part of a larger group, and is subdivided in its turn into smaller groups.

5.4. *Hypothesis construction*

A hypothesis is a provisional supposition which accounts for known facts, and can serve as a starting-point for further investigations by which it may be confirmed or refuted. An initial, usually vague, hypothesis pertaining to observational data may arise intuitively (although the observational data themselves are already unconsciously interpreted in the light of hypotheses). This initial idea can be tested against new observational data, and a new and better hypothesis can arise. From this hypothesis, sharply formulated predictions can be deduced (see the following subsection) which are now tested against precise observational data, by which they are either confirmed or refuted. In the course of an investigation, a hypothesis can be modified again, or replaced by a better one. The formulation of hypothesis is a result of creative thinking (a psychological, not a logical process); its origin is in intuition, inspiration, induction, conjecture, etc.

In order to be scientific, a hypothesis must meet a number of requirements: it must relate to a sufficiently wide range of phenomena (it must fit all the facts known at the time); it must be formulated as simply and unambiguously as possible, using clearly defined concepts; and it must be falsifiable (it must allow sufficiently precise predictions that can be tested, and subsequently confirmed or refuted).

It has been argued (Arber, 1954: 29) that hypotheses in descriptive biology are different from those in experimental biology, because they are not easily testable, and represent a way of looking at nature. Hypotheses of this kind are, for instance, those with reference to homology and relationship. It will be interesting to reconsider this view in specified cases.

5.5. *Deduction*

Deduction is the action of forming a general or particular conclusion that is completely contained in one or more given propositions. It is an objective action (in contradistinction to induction) according to particular logical rules. Deduction can be conceived in the strictly logical, or in a methodological sense; in the latter case, it includes the inference of concrete, verifiable predictions and observable facts from hypotheses. It must be borne in mind that, generally, conclusions are not drawn from a single hypothesis, but from several; predictions are based on a complete field of knowledge.

As an example of deduction, mention is made here of the systematist who identifies a specimen of a species, and verifies its conformity with the diagnosis of a species already described; he applies the deductive method, because the hypothetical model of the species is tested against the concrete case of the specimen. Another example is constituted by the systematist who has made a phenetic arrangement (according to overall similarity), and subsequently evaluates the characters by phyletic weighting; he applies the deductive method when he infers, from the hypothetical phyletic evaluation, to a greater or less degree of relationship.

5.6. *Model*

The concept of model is conceived, by different groups of investigators, in different ways. Generally, it is a simplified form or pattern of reality (the function of a model is reduction and simplification), a schematic representation designed to facilitate explanation. There must, however, be a connection between the objective reality, the model and the explanatory theory. A model should be divested of vaguenesses and superfluities, and be set down unambiguously. Models are, for instance, designed on paper, and by means of words, formulae or schemes. One can try to copy reality more or less exactly, on a smaller scale, or try to find an explanation of reality by some experiment. Any explanation, however, remains an abstraction, because only a part of reality is considered.

The description of a species, made by a systematist, is a model. It is a reduction (see subsection 5.8) of reality. A description which is really complete would be senseless because the unsurveyability continues to exist (completeness is, after all, hardly possible). A description is indeed rather realistic; it resembles its object to some extent. The systematist generally starts from one or more specimens (the description is also a model of these specimens), but generalizes in order to make the description appropriate to the whole species. A description can be too concise, or get lost in details; in both cases it is not operational, and fails as a model. In a description, those attributes can be recorded which are assumed to be important for identification and classification. An arrangement of species is based on the descriptions (specimens have too many attributes). In a first description, it will be impossible to include the complete variability of the characters; the description is functioning, in this case, also as a model for testing the range of variation.

The drawings prepared by the systematist as illustration of his description, are also functioning as models. As in the case of the description, the drawings must be applicable to the whole species. For this reason, the representation of attributes known to be accidental (such as non-essential asymmetries in Mites) should be avoided; identification is facilitated when drawings are generalized (and symmetrical). Essential asymmetries, however, such as chaotic chaetotaxies in Chelicerata, of which the asymmetry is known from experience, must indeed be represented as asymmetrical.

Reduction is still more important in the case of a diagnosis. The purpose of a diagnosis is rapid identification. A diagnostic model includes only those characters of which it is known that they are important for the discrimination of related species or groups. Also in this case, the model must be continually tested, such as in the case of the discovery of new related species, or the introduction of new related groups.

Besides descriptions, illustrations and diagnoses of species and groups, the species and groups themselves can also be understood as models. Although these models are continually tested and modified or adjusted, they keep in methodology the status of model.

The type in the sense of norm or synthesis of central values (the type of a Mammal, for instance) constitutes also a model. There are, moreover, various other kinds of types in systematics and morphology, which represent models (see Van der Hammen, 1981). The type-specimen and the type-species are more realistic and, as a model, much less appropriate.

Classifications as well as the various kinds of identification keys can also be regarded as models. Both are founded on selections of characters, which must be continually tested. Another kind of model results from the computation of similarities and differences, and the subsequent construction of diagrams of affinity.

The diagrams illustrating hypotheses with reference to evolution and phylogeny are reconstruction models; they reconstruct branching points in a dendrogram and the subsequent divergence. These models can lead to historical explanations. Several evolution models pertaining to speciation are teleological models; mention is made here of the models for the evolution of related species occurring in the same area (sympatric speciation), which are based on differences in adaptation.

5.7. *Experiment*

In an experiment a change is made in a known situation, in order to study its effect. An experiment is founded, in principle, on a comparison of results; it must be repeatable. An investigator can start from an experimental situation that is artificially created, or from a natural situation that is already existing. The concept of experiment is, however, particularly connected with artificial intervention. In an experiment, one or a few aspects of a situation are studied in particular.

An experiment is undertaken in order to test a hypothesis. Situations are created which can lead to an explanation of the problem investigated. The results obtained can lead to corrections in the formulation of the problem, and to modifications in the experiment.

Experimentation can also be an operation to discover something unknown. The investigator does not have a sharply formulated hypothesis in mind, but undertakes the experiment in order to arrive, in the course of the operation, at new hypotheses.

In systematics, the experimental method is applied to problems which cannot be solved by comparison only. This is particularly the case with closely related species or forms. A comparative morphological study can, for instance, lead to the distinction of a number of forms which differ from each other in minor details only. The hypothesis can now be introduced that the morphologically different forms are different species, and do not represent ecological races. This hypothesis can be tested by experimental hybridization and by studying the results of subsequent breeding. In the case of successful hybridization, the occurrence in nature of the intermediate forms must be studied; if these are not found, the different forms could be reproductively isolated in nature, and consequently be genuine species. In the case of parthenogenesis, hybridization is only possible in the case of arrhenotoky (where asexual and bisexual reproduction alternate). By simple breeding, one can also experimentally study the genetic variability carried in the genotype.

Another example of experimental research in systematics is constituted by breeding under experimentally changed ecological conditions, in order to be able to distinguish between phenotypical and genotypical variaton.

5.8. *Abstraction*

Abstraction is the reduction of the contents of something by separating it, in mental conception, from all kinds of less relevant details. A general concept, e.g., is formed by abstraction from observational data; those characters are selected which are considered the most essential. Nature is complicated to such a degree, that it would be unsurveyable without abstraction; by abstraction, we compensate for our restricted faculties. In the act of abstraction, the part of nature investigated by us is replaced by a simplified model. Every description is obtained by abstraction.

In biological systematics, abstraction plays a prominent part. At the species level, useful characters are separated, by abstraction, from the great mass of attributes observed. At the supraspecific level, the characters in common are obtained by abstraction; in the construction of a type (in the sense of model of a group), abstraction can play a similar part (see Van der Hammen, 1981).

5.9. *Synthesis*

Synthesis is the combining and connecting, based on experience, of separate parts or elements (such as concepts, propositions, facts) so as to make up a new complex whole which is qualitatively different from the sum of the parts. Consequently, it is not a simple summarizing of different elements, nor the indication of a central character, but the construction of something new.

Synthesis is only possible when the parts that must be combined are known. We try to start from the simplest possible elements, and to combine these into more complicated.

In systematics, synthesis plays an important part as final method. Species, genera, etc., are united into groups at the next higher level. Types (in the sense of models) can also be synthetic (see Van der Hammen, 1981).

6. Comparative study

In the action of putting two or more specimens, species or groups side by side, and considering them in connection with each other in order to mark the similarities and differences, the systematist at first compares any attributes with each other of which comparison seems to have sense. As a result of a first comparison, vague intuititve notions arise of identical structural elements which are found in different species and groups, so that they can receive the same name. Identical structural elements appear, moreover, to have a corresponding position among other elements, and these observations lead to the concept of a general plan of construction. The plan of construction is formed by abstraction. By repeated comparison of plan of construction and representatives of the group in question, and by repeated correction, the plan of construction can be developed into an abstraction model.

The larger and less homogeneous the group that is studied, the greater the chance that identical structural elements do not have the same function, or that the same function is executed by different structural elements. From these two kinds of correlations (the same form with different function, and the same function with different form) the concepts of homology and analogy have been developed. The first-mentioned concept is of paramount importance to classification.

Various attempts have been made to define the homology concept, and to find criteria for homology. Remane (1956: 28−93) has prepared a survey of three principal criteria which he arranged in order of application (an order of decreasing importance). He mentioned in the first place the criterium of identical position (particularly in the case of an equal number of structural elements with identical mutual connections). The second principal criterium is that of identical characters of the structural elements (particularly in the case of an accumulation of special characters). As a third principal criterium, Remane mentioned the presence of a series of transitions between the different forms (particularly the extremes) of a structural element.

Besides these principal criteria, Remane mentioned three auxiliary criteria for homology (particularly applicable in the case of simple structures): the presence of the structural element in a great number of closely related species; the correlation of this occurrence with other similarities; the absence of the structural element in species that are evidently not closely related.

A wide application of these criteria demonstrates that the problem of homology cannot always be solved with absolute certainty, particularly when intermediate forms are not present. For this reason a quantitative concept of homology has been developed, that permits degrees of homology (see Sneath & Sokal, 1973: 77−78). This problem is, however, not yet solved in a satisfactory way, because it has several aspects which have not been clearly distinguished: one aspect is constituted by the degree of probability of the homology in a particular case (its degree of conformability to the criteria); another aspect is constituted by such facts as important differences in ontogeny (elements which are homologous according to the above-mentioned criteria, can have developed from different embryonic material; see below).

The homology concept has repeatedly been connected with ontogenetic and phylogenetic aspects of the structural elements. Remane has demonstrated that, in a great many cases, structural elements which are undoubtedly homologous, can have a different embryonic origin, as well as a different subsequent development. That does not alter the fact that, in other cases, characters from ontogeny can be of great value, particularly in connection with the criterium of identical special characters.

Since Darwin, it has repeatedly been claimed that structural elements are homologous when they have a common phylogenetic origin. This criterium is, however, founded on circular reasoning: phylogenies are first worked out on the base of homologies of primitive and derived character states, so that homologies cannot be subsequently defined in terms of phylogeny. The evolutionary explanation of homology ('the homology of an element in two or more species is due to inheritance from a common ancestor') is, in fact, a hypothesis.

When, in the course of a comparative study, kinds of attributes other than morphological are compared, it must be examined whether the homology concept can also be used in these cases, and which criteria can be applied. Little progress has been made with the development of a general theory of comparison (see Woodger, 1945; Cain & Harrison, 1958; Cain, 1968). It must be assumed that, in the course of a general comparison, those characters (and their connections in space and time) can be particularly considered for comparison, that have a similar position among other corresponding characters (the same topological relations).

The homology concept has been thoroughly studied in ethology (see Baerends, 1970: 160–163), and the following criteria are used there: similarity in the stereotyped form of a movement; causation by similar internal and external stimuli; similarity in the position of a particular action in the total behavioural pattern; and the presence of transitions between extremes.

In several other branches of biology, satisfactory criteria have not been developed. In biochemistry, the criterium of similar biosynthesis has been used; the difficulties in this branch are demonstrated by the widespread occurrence of hemoglobin (which is found in representatives of widely separated groups like Mollusca, Annelida, Insecta, Vertebrata, and Angiospermae). Fitch (1970) has tried to distinguish between homologous and analogous proteins; he wrongly defined the homology concept in terms of phylogeny, reconstructed phylogeny hypothetically by 'intelligent guessing', and drew conclusions pertaining to homology or analogy by deduction from this hypothesis. In molecular biology, the concept of DNA-homology is used. This concept pertains to the degree of similarity of chromosomes, homology being expressed in percentages of the total length. Evidently, homology is defined here in a different way.

For a general comparison, the topological criterium (pertaining to the similar position, in space and time, of an element among other elements) is, undoubtedly, the most important. Evidently, the application of the homology concept to comparative studies is most advanced in morphology (see also Patterson, 1982). In all cases where the functional aspect of a character is dominating, the possibility of analogy must be seriously considered.

As mentioned in subsection 5.2, it is really necessary, for an objective comparison, that similarities and differences are measured. Methods for coding and scaling of characters, and for computation of measures of resemblance, have been developed particularly in numerical taxonomy (see Sneath & Sokal, 1973). In any case where characters are logically correlated (such as the presence of a pigment and the observation of a particular colour), these are coded as a single character. Characters which are empirically correlated (such as the characters in common of a group) are coded separately.

7. The species-concept

The origin of the species-concept is in folk taxonomy, where it refers to similar specimens which can be indicated by the same name. For ages, many people had unconsciously applied this common unwritten definition, before attempts were made to define the concept in a more scientific way.

The species-concept of Aristotle is closely related to his theory of form and matter: different specimens of the same species have the same immanent form (now to be understood in the sense of genotype). In the species-concept of Linnaeus (see Engel, 1953), much importance is attached to the self-perpetuating power of species, and to the Creation of the first specimens of each species, although Linnaeus' opinions on the constancy of species changed in the course of time (in his later works he supposed that new species could arise by hybridization). Modern systematists have tried to extend the definition of the species-concept by considering also various other aspects.

There are several properties by which the biological species transcends the simple interpretation as a class of objects. These properties pertain to: the synbiological aspect of the species (the mutual relations between representatives of a population, and the relations between that population and populations of other species occurring in the same area); the genetical aspect of the species (the species as a genetical unit, which derives its reality from the shared information present in the gene pool); and the evolutionary aspect (the species as part of a phylogenetic line).

Several modern definitions of the species appeared to be unsatisfactory because they can only be applied to particular cases. This is, for instance, the case with the definition of the so-called biological species ('species are groups of interbreeding natural populations that are reproductively isolated from other such groups'; see Mayr, 1969: 26), which cannot be applied to the great majority of species (not only because of the occurrence of parthenogenesis, but because of the practical difficulties with breeding experiments, particularly in the case of exotic material described by a museum systematist). In one of the most recent definitions, the species is simply defined as 'the smallest detected samples of self-perpetuating organisms that have unique sets of characters' (Nelson & Platnick, 1981: 12).

Ghiselin (1966: 208–209; 1975) recently argued that species (in the sense of taxa) may be considered individuals in the logical sense of particular thing (the species category, however, is a class). The species name, consequently, is a proper name, and not the universal name of a class. The constituent organisms of species (in the logical sense of individuals) are parts, not members. Hull (1976) subsequently developed this view by positing that the species, in the logical sense of individual, is a unit of evolution, a 'chunk in the genealogical nexus', which exists between two successive speciation events (see also: Reed, 1979; Mishler & Donoghue, 1983; Kitts, 1983).

In taxonomic practice, the species is introduced as a model (in the sense of simplified pattern of reality) of the smallest taxonomic unit; this model is obtained by abstraction, and is represented by the description. It is increasingly perfected when one proceeds, from the species of the museum systematist (the morphospecies, in the sense of Cain, 1954), through experimental testing, in the direction of the biological species or the agamospecies (in the sense of Cain, 1954). The model implies several hypotheses which, generally, are not explicitly mentioned. Some of these hypotheses can be the following.

a. The species is self-perpetuating, and similar specimens can subsequently be found by any collector. Generally, this hypothesis will be repeatedly confirmed later on (except when species become extinct).

b. The description sufficiently characterizes the species. Subsequent observations may, however, lead to corrections in the description; the range of variability may, for instance, be wider than originally supposed, or characters supposed to be specific may prove to be secondary sexual.

c. The species occupies a geographical area larger than the original type-locality. Many observations will be required before this hypothesis can be reconstructed, in referring to a distinctly limited geographical area.

d. In the case of markedly different phena, constituting one species, the phena are first hypothetically regarded as conspecific. This hypothesis can be confirmed by studies of the reproductive behaviour and by breeding (circumstances permitting this experiment), and by evidence from closely related species.

e. The species as described is the result of evolution, and will exist in a limited space of time. Fossil specimens of the species may be found, during paleontological research, by going back as far as the speciation event that gave rise to it. Future evolutionary changes can be predicted in the case of so-called vertitional changes (see Van der Hammen, 1981a), presupposed that the evolutionary program of the group in question is already known by a comparative study of related species. In the case of species of the superfamily Nothroidea

(Oribatid mites), e.g., many predictions can be made as to the future evolution of chaetotaxy.

8. Biological classification

Biological classification involves the placing of similar species and groups (genera, families, orders, classes, etc.) in higher units, separated from groups of the same kind by distinct discontinuities; these higher units, which are defined by the possession of characters in common, are subsequently ordered, ranked and named.

As mentioned in subsection 5.3, an intuitive classification, based on unformulated experience, is found in primitive societies and, generally, in simple, untrained people. Biological systematics has tried to improve this approach, on the one hand by adopting methods from philosophy (it developed the method of classifying according to genus and species, by introducing a hierarchical system extending from species to phylum and regnum), on the other hand by developing classification into a branch of science (by classifying according to hypothetical phylogenetic relationship).

A supraspecific taxon is constituted by a group of related species (or, in the case of monotypic taxa, a single species); it is separated from groups of the same kind by discontinuities. A supraspecific category is a class of taxa of the same rank. In a hierarchical classification (which is the result of comparative studies; it is based on a hierarchy of basic structural patterns), each group forms part of a larger group and is subdivided in its turn into smaller groups; ranking can at first be arbitrary. In the hierarchy of the natural system, the genus occupies a special place; it is, in many groups, the most 'natural' of the supraspecific taxa (it can often be recognized without detailed study), and its name is included in the species name. The 'naturalness' of supraspecific taxa (i.e., the general likeness of all of their species to a single representative type) becomes increasingly less evident at higher levels of the taxonomic hierarchy, and can become problematic at the supraordinal levels.

The 'naturalness' of supraspecific taxa is attributable to relationship and, probably, often also to the filling of a well-defined new adaptive zone by the common ancestors. Biological systematics has made attempts to classify according to these hypothetical relationships. It is supposed that evolution is reflected by the hierarchy of the plans of construction, and that the characters are older (in an evolutionary sense) as the hierarchical level is higher. In many cases, a sequence of taxa in the hierarchy of a classification is supposed to constitute a phylogenetic line. All characters of the taxonomic hierarchy are supposed to be included in the genotypic program.

The final presentation of a classification depends on the ideal we have in view (critics often forget to analyse this important point). In the procedure of classifying, two phases can be distinguished: sorting and grouping according

178

to overall similarity, and the modification of this grouping by phyletic weighting of characters (interpretation of character states, corrections for convergence) and the subsequent attribution of a particular phylogenetic value to some attributes. In a phenetic classification particular attention is paid to overall similarity, in an evolutionary classification to evolutionary divergence, and in a phylogenetic classification to branching points. Differences of opinion on the average size of taxa (splitting or lumping) are also of influence on the final presentation.

A biological (i.e., evolutionary or phylogenetic) classification must be regarded as a model of relationship, and, in its hierarchical arrangement, as a model of phylogeny and the successive levels of new potentialities. As such it implies several hypotheses, of which the following are mentioned here.

a. The species are placed near to or far from each other according to the measure of relationship.

b. The character states are correctly interpreted, which implies that this interpretation is in accordance with the phylogenetic sequence of manifestation.

c. Hierarchical ranking reflects phylogeny; characters of the highest levels of classification are indeed those of the oldest levels of the potentialities.

For the construction of an adequate model, as many data as possible should be available, particularly those pertaining to the interpretation of character states, and the interpretation of levels of adaptation. These data together constitute the standard of classification, so that all classifications based on the same standard will be comparable. A model which comes up to these requirements, will have predictive value: with a correct choice of characters, the discovery of new characters can be expected. In the subdivision of the Chelicerata into subclasses (Van der Hammen, 1977a), for instance, the hypothesis was introduced that the segmentation and articulation of the legs referred to one of the oldest levels of adaptation, and, consequently, to one of the highest levels of classification. This hypothesis was subsequently confirmed by the discovery of several very interesting new characters (Van der Hammen, 1979a, 1982, 1985, 1986a; 1986b: 224–228, Fig. 3).

In view of the many different (and sometimes highly artificial) classifications of the representatives of one and the same taxonomic group, which can be prepared according to current taxonomic methods, it may be wondered if the rational reconstruction of systematic activity is perhaps still incomplete. Current taxonomic methods tend indeed to develop atomistic views in which the organism as a functional whole has virtually no place; elements can, for instance, be left unconsidered because of their supposed primitiveness, although their new role in an architectural whole has an advanced character (as in the case of chelicerate epimera). In the structuralistic model recently introduced by me (see Van der Hammen, 1981, 1985a, 1986), which is based on the hypothesis of a hierarchy of transformation systems, some of the imperfections of current methods are corrected and supplemented. In this model attention is paid, for instance, to the evolutionary potentialities of the genotype; branching points are supposed to represent subdivisions into different evolutionary programs, whilst these programs can subsequently manifest themselves separately by parallel evolution. It is pointed out that seeming similarities in the change of character states can be the result of the activity of different regulatory mechanisms, by which the changes are not homologous (see Van der Hammen, 1986). Attention is further paid to the hierarchical character of the model and the necessity of a hypothetical explanation of the association between character (or character complex) and hierarchic value. It is also demonstrated (see Van der Hammen, 1986b) that the manifestation of important characters, particularly at the higher levels in the hierarchy of transformation systems, can subsequently be superseded by the manifestation of other characters. Evidently, taxonomic methodology is still incompletely reconstructed and developed.

References

Arber, A., 1954. The mind and the eye. – Cambridge (University Press). [I have used the second edition, Cambridge, 1964: xii + 146 p.]

Aubert, H. & Fr. Wimmer, 1868. Aristoteles Thierkunde. Kritisch-berichtigter Text mit Deutscher Übersetzung, sachlicher und sprachlicher Erklärung und vollständigem Index. – Leipzig, vol. 1: viii + 543 p.

Baerends, G.P., 1970. De bruikbaarheid van gedragskenmerken voor de zoölogische systematiek. – In: K.H. Voous (ed.), Biosystematiek (Wageningen): 145–179, Figs. 1–10.

Balme, D.M., 1962. Genos and eidos in Aristotle's biology. – The Classical Quarterly (N.S.), 12: 81–98. [A German translation is included in: G.A. Seeck (ed.), Die Naturphilosophie des Aristoteles (Darmstadt, Wissenschaftliche Buchgesellschaft, Welt der Forschung 225): 139–171.]

Cain, A.J., 1954. Animal species and their evolution. – London (Hutchinson). [I have used the second edition, reprinted 1968: 190 p., 5 figs.]

Cain, A.J., 1968. The assessment of new types of character in taxonomy. – In: J.G. Hawkes (ed.), Chemotaxonomy and serotaxonomy (The Systematics Association, spec. vol. 2): 229–234.

Cain, A.J. & G.A. Harrison, 1958. An analysis of the taxonomist's judgement of affinity. – Proc. Zool. Soc. London: 131: 85–98, Fig. 1, Tab. 1–4.

Camp, W.H. & C.L. Gilly, 1943. The structure and origin of species. – Brittonia, New York 4: 323–385.

Candolle, A.P. de, 1813. Théorie élémentaire de la botanique, ou exposition des principes de la classification naturelle et de l'art de décrire et d'étudier les végétaux. – Paris: viii + 500 + 27 p.

Conklin, H.C., 1962. Lexicographical treatment of folk taxonomies. – Int. Journ. Amer. Linguist. 28: 119–141, Figs. 1–3.

Engel, H., 1953. The species concept of Linnaeus. – Arch. Int. Hist. Sci. 23/24: 249–259.

Fitch, W.M., 1970. Distinguishing homologous from analogous proteins. – Syst. Zool. 19: 99–113.

Gertsch, W.J., 1964. A review of the genus Hypochilus and a description of a new species from Colorado (Araneae, Hypochilidae). – Amer. Mus. Novit. 2203: 1–14, Figs. 1–11.

Ghiselin, M.T., 1966. On psychologism in the logic of taxonomic controversies. – Syst. Zool. 15: 207–215.

Ghiselin, M.T., 1975. A radical solution to the species problem. – Syst. Zool. 23: 536–544.

Hammen, L. van der, 1971. Over de notatie van vogelgeluiden. – De Levende Natuur 74: 128–134, Figs. 1–4.

180

Hammen, L. van der, 1972. Kleurwaarnemingen aan planten. – De Levende Natuur 75: 200–209, Figs. 1–6.

Hammen, L. van der, 1974. Botanische geurbeschrijvingen. – De Levende Natuur 76: 238–244.

Hammen, L. van der, 1975. Aanvullende gegevens over geuren, de reuk en de geurbeschrijving. – De Levende Natuur 77: 196–203, Figs. 1–5.

Hammen, L. van der, 1975a. Leonardo da Vinci's aantekeningen over de vogelvlucht: Een inleiding tot het waarnemen van beweging. – Museologia 4: 28–38, Figs. 1–7.

Hammen, L. van der, 1975b. Op de tast door het plantenrijk. – De Levende Natuur 78: 192–202, Figs. 1–5.

Hammen, L. van der, 1977. Planten proeven. – De Levende Natuur 80: 49–55, Figs. 1–2.

Hammen, L. van der, 1977a. A new classification of Chelicerata. – Zool. Med. Leiden 51: 307–319, Figs. 1, Tab. 1–3.

Hammen, L. van der, 1978. Waarnemingen met betrekking tot licht en donker. – De Levende Natuur 81: 21–27, Figs. 1–3.

Hammen, L. van der, 1978a. The evolution of the chelicerate life-cycle. – Acta Biotheoretica 27: 44–60, Figs. 1–6. [Cf. essay III.]

Hammen, L. van der, 1979. Inleiding tot het waarnemen van vorm. – De Levende Natuur 81: 225–233, Figs. 1–7.

Hammen, L. van der, 1979a. Comparative studies in Chelicerata I. The Cryptognomae (Ricinulei, Architarbi and Anactinotrichida). – Zool. Verh. Leiden 174: 1–62, Figs. 1–31.

Hammen, L. van der, 1981. Type-concept, higher classification and evolution. – Acta Biotheoretica 30: 3–48, Figs. 1–7. [Cf. essay V.]

Hammen, L. van der, 1981a. Numerical changes and evolution in Actinotrichid mites (Chelicerata). – Zool. Verh. Leiden 182: 1–47, Figs. 1–13. [Cf. essay II.]

Hammen, L. van der, 1982. Comparative studies in Chelicerata II. Epimerata (Palpigradi and Actinotrichida). – Zool. Verh. Leiden 196: 1–70, Figs. 1–31.

Hammen, L. van der, 1983. Unfoldment and manifestation: The natural philosophy of evolution. – Acta Biotheoretica 32: 179–193. [Cf. essay I.]

Hammen, L. van der, 1985. Comparative studies in Chelicerata III. Opilionida. – Zool. Verh. Leiden 220: 1–60, Figs. 1–34.

Hammen, L. van der, 1985a. A structuralist approach in the study of evolution and classification. – Zool. Med. Leiden 59: 391–409, Figs. 1–5. [Cf. essay VI.]

Hammen, L. van der, 1986. On some aspects of parallel evolution in Chelicerata. – Acta Biotheoretica 35: 15–37, Figs. 1–9. [Cf. essay IV.]

Hammen, L. van der, 1986a. Comparative studies in Chelicerata IV. Apatellata, Arachnida, Scorpionida, Xiphosura. – Zool. Verh. Leiden 226: 1–52, Figs. 1–23.

Hammen, L. van der, 1986b. Acarological and arachnological notes. – Zool. Med. Leiden 60: 217–230, Figs. 1–3.

Hammen, L. van der, 1986c. Some notes on taxonomic methodology. – Zool. Meded. Leiden 60: 231–256.

Hennig, W., 1950. Grundzüge einer Theorie der phylogenetischen Systematik. – Berlin: viii + 370 p.

Hennig, W., 1966. Phylogenetic systematics. – Urbana, Chicago, London (University of Illinois Press): viii + 263 p., 69 figs.

Hull, D.L., 1976. Are species really individuals? – Syst. Zool. 25: 174–191.

Kälin, J., 1941. Ganzheitliche Morphologie und Homologie. – Freiburg: 41 p, 3 figs.

Kitts, D.B., 1983. Can baptism alone save a species? – Syst. Zool. 32: 27–33.

Mayr, E., 1969. Principles of systematic zoology. – New York, etc. (McGraw-Hill Book Company): xiv + 428 p., figs., tab.

Mayr, E., E.G. Linsley & R.L. Usinger, 1953. Methods and principles of systematic zoology. – New York, etc. (McGraw-Hill Book Company): x + 328 p., figs., tab.

Mishler, B.D. & M.J. Donoghue, 1983. Species concepts: a case for pluralism. – Syst. Zool. 31: 491–503.

Nelson, G. & N. Platnick, 1981. Systematics and biogeography; cladistics and vicariance. – New York (Columbia University Press): xi + 567 p.

Patterson, C., 1982. Morphological characters and homology. – In: K.A. Josey & A.E. Friday (eds.), Problems of phylogenetic reconstruction (London, Academic Press): 21–74.

Reed, E.S., 1979. The role of symmetry in Ghiselin's 'radical solution of the species problem'. – Syst. Zool. 28: 71–78.

Remane, A., 1952. Die Grundlagen des natürlichen Systems, der vergleichenden Anatomie und der Phylogenetik. – Leipzig: 400 p.

Remane, A., 1956. Die Grundlagen des natürlichen Systems, der vergleichenden Anatomie und der Phylogenetik. Theoretische Morphologie und Systematik I. – Leipzig (second edition): vi + 364 p., 82 figs. [I have used the reprint edition, Koenigstein-Taunus (Otto Koeltz), 1971.]

Simpson, G.G., 1961. Principles of animal taxonomy. – New York, London (Columbia University Press): xiv + 247 p., 30 figs.

Sneath, P.H.A. & R.R. Sokal, 1962. Numerical taxonomy. – Nature, London 193: 855–860, Figs. 1–5.

Sneath, P.H.A. & R.R. Sokal, 1973. Numerical taxonomy. The principles and practice of numerical classification. – San Francisco (W.H. Freeman and Company): xvi + 573 p., figs., tab.

Sokal, R.R. & P.H.A. Sneath, 1963. Principles of numerical taxonomy. – San Francisco, London (W.H. Freeman and Company): 359 p.

Vachon, M., 1952. Étude sur les Scorpions. – Alger (Institut Pasteur d'Algérie): 482 p., 697 figs.

Wiley, E.O., 1981. Phylogenetics: The theory and practice of phylogenetic systematics. – New York (John Wiley): xv + 439 p., figs.

Woodger, J.H., 1945. On biological transformations. – In: W.E. Le Gros Clark & P.B. Medawar (eds.), Growth an form. Essays presented to D'Arcy Wentworth Thompson (Oxford): 94–120.